Advances in Intelligent Systems and Computing

Volume 305

Series editor

Janusz Kacprzyk, Polish Academy of S

e-mail: kacprzyk@ibspan.waw.pl

About this Series

The series "Advances in Intelligent Systems and Computing" contains publications on theory, applications, and design methods of Intelligent Systems and Intelligent Computing. Virtually all disciplines such as engineering, natural sciences, computer and information science, ICT, economics, business, e-commerce, environment, healthcare, life science are covered. The list of topics spans all the areas of modern intelligent systems and computing.

The publications within "Advances in Intelligent Systems and Computing" are primarily textbooks and proceedings of important conferences, symposia and congresses. They cover significant recent developments in the field, both of a foundational and applicable character. An important characteristic feature of the series is the short publication time and world-wide distribution. This permits a rapid and broad dissemination of research results.

Advisory Board

Chairman

Nikhil R. Pal, Indian Statistical Institute, Kolkata, India
e-mail: nikhil@isical.ac.in

Members

Rafael Bello, Universidad Central "Marta Abreu" de Las Villas, Santa Clara, Cuba
e-mail: rbellop@uclv.edu.cu

Emilio S. Corchado, University of Salamanca, Salamanca, Spain
e-mail: escorchado@usal.es

Hani Hagras, University of Essex, Colchester, UK
e-mail: hani@essex.ac.uk

László T. Kóczy, Széchenyi István University, Győr, Hungary
e-mail: koczy@sze.hu

Vladik Kreinovich, University of Texas at El Paso, El Paso, USA
e-mail: vladik@utep.edu

Chin-Teng Lin, National Chiao Tung University, Hsinchu, Taiwan
e-mail: ctlin@mail.nctu.edu.tw

Jie Lu, University of Technology, Sydney, Australia
e-mail: Jie.Lu@uts.edu.au

Patricia Melin, Tijuana Institute of Technology, Tijuana, Mexico
e-mail: epmelin@hafsamx.org

Nadia Nedjah, State University of Rio de Janeiro, Rio de Janeiro, Brazil
e-mail: nadia@eng.uerj.br

Ngoc Thanh Nguyen, Wroclaw University of Technology, Wroclaw, Poland
e-mail: Ngoc-Thanh.Nguyen@pwr.edu.pl

Jun Wang, The Chinese University of Hong Kong, Shatin, Hong Kong
e-mail: jwang@mae.cuhk.edu.hk

More information about this series at http://www.springer.com/series/11156

Rituparna Chaki · Khalid Saeed
Sankhayan Choudhury · Nabendu Chaki
Editors

Applied Computation and Security Systems

Volume Two

Springer

Editors
Rituparna Chaki
A.K. Choudhury School of Information
    Technology
University of Calcutta
Kolkata, West Bengal
India

Sankhayan Choudhury
Nabendu Chaki
Department of Computer Science
    and Engineering
University of Calcutta
Kolkata, West Bengal
India

Khalid Saeed
Faculty of Physics and Applied Computer
    Sciences
AGH University of Science and Technology
Cracow
Poland

ISSN 2194-5357   ISSN 2194-5365 (electronic)
ISBN 978-81-322-1987-3   ISBN 978-81-322-1988-0 (eBook)
DOI 10.1007/978-81-322-1988-0

Library of Congress Control Number: 2014947644

Springer New Delhi Heidelberg New York Dordrecht London

Printed on acid-free paper

Springer is part of Springer Science+Business Media (www.springer.com)

Preface

The First International Doctoral Symposium on Applied Computation and Security Systems (ACSS 2014) took place during Apr 18–20, 2014 in Kolkata, India. This symposium is aimed to facilitate the Ph.D. students to present and discuss their research work leading towards high-quality dissertation. This symposium will provide a friendly and supportive environment for doctoral students to present and discuss their work both with their peers and with a panel of distinguished experts. ACSS Doctoral Symposium allowed researchers working in different fields of computer science such as Image processing, Remote Healthcare, Biometrics, Pattern Recognition, Embedded Systems, Data Mining, Software Engineering, Networking, and Network Security. The symposium evolved as a joint venture between two collaborative universities: the University of Calcutta, India, and the AGH University of Science and Technology, Poland.

The program committee members of ACSS 2014 were instrumental in disseminating the objectives of the symposium among the scholars and faculty members in a very short time. This resulted in a large number of submissions from Ph.D. scholars from India and abroad. These papers underwent a minute and detailed blind-review process with voluntary participation of the committee members and external expert reviewers. The metrics for reviewing the papers had been mainly the novelty of the contributions, technical content, organization, and clarity in presentation. The entire process of initial paper submission, review, and acceptance were done electronically. The hard work done by the Organizing and Technical Program Committees led to a superb technical program for the symposium. The ACSS 2014 resulted in high-impact and highly interactive presentations by the doctoral students.

The Technical Program Committee for the symposium has selected only 25 papers for publication out of a total 70 submissions. Session chairs were entrusted with the responsibility of submitting feedbacks for improvements of the papers presented. The symposium proceeding has been organized as a collection of papers, which were presented and then modified as per reviewer's and session chair's comments. This has helped the scholars to further improve their contributions.

We would like to take this opportunity to thank all the members of the Technical Program Committee and the external reviewers for their excellent and time-bound review works. We especially thank Prof. Indranil Sengupta of IIT, Kharagpur for his suggestions towards designing the Technical Program for ACSS-2014. We thank all our sponsors who have come forward towards organization of this symposium. These include Tata Consultancy Services (TCS), Springer India, ACM India, M/s Business Brio, M/s Enixs. We appreciate the initiative and support from Mr. Aninda Bose and Ms. Kamiya Khatter his colleagues in Springer for their strong support towards publishing this post-symposium book in the series "Advances in Intelligent Systems and Computing." Last, but not the least, we thank all the authors without whom the symposium would not have reached up to this standard.

On behalf of the editorial team of ACSS 2014, we sincerely hope that the different chapters of this book will be beneficial to all its readers and motivate them towards further research.

Rituparna Chaki
Khalid Saeed
Sankhayan Choudhury
Nabendu Chaki

Contents

About the Editors

Rituparna Chaki is an Associate Professor in the A.K. Choudhury School of Information Technology, University of Calcutta, India since June 2013. She joined the academia as faculty member in the West Bengal University of Technology in 2005. Before that she has served under Government of India in maintaining industrial production database. Rituparna has done her Ph.D. from Jadavpur University in 2002. She has been associated in organizing many conferences in India and abroad as Program Chair, OC Chair, or as member of Technical Program Committee. She has published more than 60 research papers in reputed journals and peer-reviewed conference proceedings. Her research interest is primarily in Ad-hoc networking and its security. She is a professional member of IEEE and ACM.

Khalid Saeed received the B.Sc. Degree in Electrical and Electronics Engineering from Baghdad University in 1976, the M.Sc. and Ph.D. Degrees from Wroclaw University of Technology, in Poland in 1978 and 1981, respectively. He received his D.Sc. Degree (Habilitation) in Computer Science from Polish Academy of Sciences in Warsaw in 2007. He is a Professor of Computer Science with AGH University of Science and Technology in Poland. He has published more than 200 publications—edited 23 books, Journals and Conference Proceedings, 8 text and reference books. He supervised more than 110 M.Sc. and 12 Ph.D. theses. His areas of interest are Biometrics, Image Analysis, and Processing and Computer Information Systems. He gave 39 invited lectures and keynotes in different universities in Europe, China, India, South Korea, and Japan. The talks were on Biometric Image Processing and Analysis. He received about 18 academic awards. Khalid Saeed is a member of more than 15 editorial boards of international journals and conferences. He is an IEEE Senior Member and has been selected as IEEE Distinguished Speaker for 2011–2016. Khalid Saeed is the Editor-in-Chief of International Journal of Biometrics with Inderscience Publishers.

Sankhayan Choudhury is Associate Professor in the Department Computer Science and Engineering, University of Calcutta, India. Currently, he is head of this department. Moreover, he is Co-ordinator of TEQIP-II, University of Calcutta. Dr. Choudhury has obtained his B.Sc. (Hons.) in Mathematics under University of Calcutta. Thereafter he has obtained B.Tech. and M.Tech in Computer Science and Engineering from University of Calcutta. He has completed Ph.D. from Jadavpur University, India in 2006. His research interests include Mobile Computing, Networking, Sensor Networking, Cloud Computing, etc. Besides authoring a book, Dr. Choudhury has published close to 50 peer-reviewed papers in international journals and conference proceedings. He has also served in the Program Committees of several international conferences and has also chaired the Program and Organizing Committees of a few. Dr. Choudhury is a professional member of ACM and an executive committee member for the local ACM professional chapter in Kolkata, India.

Nabendu Chaki is a Senior Member of IEEE and an Associate Professor in the Department Computer Science and Engineering, University of Calcutta, India. Besides editing several volumes in Springer in LNCS and other series, Nabendu has authored three textbooks with reputed publishers like Taylor and Francis (CRC Press), Pearson Education, etc. Dr. Chaki has published more than 120 refereed research papers in Journals and International conferences. His areas of research interests include image processing, distributed systems, and network security. Dr. Chaki has also served as a Research Assistant Professor in the Ph.D. program in Software Engineering in U.S. Naval Postgraduate School, Monterey, CA. He is a visiting faculty member for many universities including the University of Ca'Foscari, Venice, Italy. Dr. Chaki has contributed in SWEBOK v3 of the IEEE Computer Society as a Knowledge Area Editor for Mathematical Foundations. Besides being in the editorial board of Springer and many international journals, he has also served in the committees of more than 50 international conferences. He is the founding Chapter Chair for ACM Professional Chapter in Kolkata, India since January 2014.

Part I
Software Engineering

Non-functional Property Aware Brokerage Approach for Cloud Service Discovery

Adrija Bhattacharya and Sankhayan Choudhury

Abstract In the fast growing service-oriented domain, cloud computing becomes the focal issue of the current research initiatives. Increase in cloud services, providers, consumers and their requirements demands efficient handling of the inherent complexity in a cloud environment. Cloud service broker (CSB) is one of such initiative. Research on CSB, an inter-mediatory, has opened an unexplored domain of service provisioning techniques. Seamless service provisioning with better QoS (such as cost and time) is one of the major challenges in CSB design. In this paper, an attempt has been made to propose a framework for facilitating service provisioning techniques within CSB. The non-functional parameter (NFP) along with functional one plays important role in service discovery from a set of offered services. The service discovery process based only on functionality may lead to an infeasible, hence unaccepted solution to a consumer. The meta-model, a proposed component within CSB, in the form of a lattice is introduced for speed up and to select more relevant set of services satisfying requirements of the consumers. The lattice grabs service information for all possible NFP combinations in a structured way. Different lattices with respect to each cloud layers are constructed independently. These lattices are exploited for finding the most relevant services. Moreover, it is expected to behave better in terms of search time.

Keywords Service discovery · Lattice · Cloud broker

A. Bhattacharya (✉) · S. Choudhury
Department of Computer Science and Engineering, University of Calcutta, Kolkata, India
e-mail: adrija.bhattacharya@gmail.com

S. Choudhury
e-mail: sankhayan@gmail.com

© Springer India 2015   3
R. Chaki et al. (eds.), *Applied Computation and Security Systems*, Advances in Intelligent
Systems and Computing 305, DOI 10.1007/978-81-322-1988-0_1

1 Introduction

In the fast growing service-oriented domain, cloud computing becomes the focal issue of the current research initiatives. The cloud services are very much different [1] and efficient with respect to Web services. Cloud actually offers software (computational), platform (behavioural) and infrastructural data resources as service from remote sources on demand. It consists of three types of models: Software as a Service (SaaS), Platform as a service (PaaS) and Infrastructure as a Service (IaaS) [2]. These three types of cloud services are often combined in different manner to satisfy user requests.

The increase in different service offering as well as numbers of service providers has put up a new challenge to the cloud researchers as these poses enormous different service provisioning with similar functionality but varying performance. The Service Level Agreement (SLA) is a mediator document that has to be satisfied for providing service to any consumer. In cloud computing paradigm, a consumer may be an end user or another service provider. In spite of developing and maintaining SLAs, cloud services often need mediation for coordinating SLAs, consumer management, reporting, pricing and accounting, etc. In this context, the cloud service broker (CSB) is proposed as a solution [3].

Cloud service provisioning through brokers is an open problem area of research. A broker may be considered as an upgradation on classical multi-source integrator used in distributed environments [4]. The service with similar functionality in a cloud environment is differentiated by performance attributes such as granularity, outcome, governance and control which works for betterment of cloud service systems with the help of CSBs. Thus, service provisioning techniques offered by a broker becomes a challenge.

An overall description of a broker is depicted in Fig. 1. There exists no such standardized architecture of CSB. There are three major aspects, namely aggregation, integration and customization, based on which a CSB can be formed. Aggregation composes two or more service from one or more providers and then delivers it to the user or any other service provider. Appstore is a typical example of service aggregation through broker. Integration is another role that leads to substantially new values to community management in cloud domain [5]. Gmail contact list is the real-life example of integration by cloud broker. In customization, newer functionalities are added to improve the existing service functionality. The offered functionality of a broker can be customized in a flexible way based on a specific application domain. Thus, automatic and technically advanced methodologies are needed to enhance the strength of CSBs.

In Fig. 1, the component, marked in yellow, is the area of interest. In this paper, our objective is to provide an enhanced service discovery mechanism with the help of proposed meta-model. The said meta-model holds the detailed service information and can be used for finding most relevant services in service provisioning. Proposed enhanced service discovery mechanism is expected to behave in a more

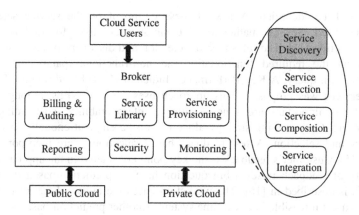

Fig. 1 Cloud service broker

efficient way in terms of overhead (time). The penalty is to be paid in terms of time needed for model creation. This is insignificant as it is created offline, but can be used several times to satisfy user query.

2 Related Work

This section gives a brief review on service provisioning role of cloud broker. In recent time, a significant upgradation on existing cloud service architecture has been done. A collaborative cloud market model [6] is proposed for collaboration among service clouds by means of resource sharing. Guaranteed QoS [7] from consumer side is considered for the first time, but this issue demands higher overhead in terms of time. In another research work, a ranking system for cloud service composition and goal-based method based on proposed SLA template for discovering proper monitoring services have been incorporated [2]. Cloud service provisioning varies in many ways with respect to service provisioning policies in Web. Lack of standardized representation of the cloud providers' criteria and varying format and content of SLAs made the existing Web service discovery and selection policies infeasible to implement in cloud domain [1]. The tightly coupled nature of services of a cloud often lacks flexibility, and as a result, one may not be able to club services offered by different providers. Like, if a user has chosen Amazon S3 cloud storage, then he is compelled to use Amazon provider's services [8]. Besides that challenge, due to complex relationship among cloud service providers, the existing service composition policies also failed to accumulate required cloud services within a composed single service. References [3, 9] have mentioned the required functionalities of Cloud brokers such as service monitoring and service aggregation.

The work of service description is the foremost to remember. A formal and detailed functional description about CSB and its role can be primarily identified in [10]. Through the review work, an interesting taxonomy among the existing works has been identified. A few of these comment on the service selection techniques used. Several mathematical techniques are used for ranking-based service selection [11]. The relevance of selection and discovery of cloud services depends on the required functionality and non-functional parameters (NFP) specification in query. A Key Performance Indicator (KPI)-based ranking [12] of cloud services was proposed. This approach uses a set of NFPs, among which some of the important are time, accuracy, interoperability, cost, reliability, usability, etc. Only ranking strategy alone cannot be efficient enough for service discovery and selection. A few feedback-based selections were incorporated in [13, 14, 15]. Customer's feedback is the key point considered in the approach. The reliability of feedback raises a big question here. A prediction-based discovery approach is described in [16]. The prediction bases are huge volume of previous data. It is quiet infeasible in a real time system. Another prediction-based selection is in [17]. But all of these are domain-specific applications. None of these takes all possible NFPs into account, most of these dealing with the parameter "cost". Obviously, it is the most important one, but it alone cannot be taken as the key criteria for service selection.

Often the broker has a role in service decision-making and resource optimization. A set of works on this exists; for example, [18, 19] can be thought as pioneer in the domain. But the optimization often leads to cost optimization which is not the only need. Another cost optimization-based selection is done in [20]. It uses two-way optimization, but this is a very problem-specific method and can only be applied in feature placement problem.

The above discussion, especially on service provisioning, establishes the need for a better service discovery mechanism that should consider the functional as well as non-functional attributes as requested by consumers in a cloud environment. Most of the existing techniques are not considering all relevant NFPs used in the query, and as a result, it may generate an infeasible solution for the consumer. The scope of the work offers a meta-model that holds service information in the contexts all possible NFPs such that it can be used for better service provisioning as a whole.

3 Proposed Solution

The service provisioning in the area of cloud computing involves three main entities in a dialogue session. Consumer agent and provider agent both interact through third entity, broker agent or a set of broker agents. Brokers hold information about multiple cloud providers and their services. Each service (SaaS, PaaS, IaaS, etc.) has some non-functional specifications from provider's end. Consumer's requirement has two parts one is functional and another is non-functional requirements. Functional requirements of a query may be satisfied by either a single service (any of SaaS, PaaS, IaaS, etc.) or a set of multiple services (any combination of SaaS, PaaS, IaaS, etc.). In both of the cases, a minimum required non-functionality (specified by user) has to be satisfied. The reason behind is that in multi-cloud environment, information about all the service offerings is large in size and impossible to accumulate at single point from where a consumer can look for. In the process of searching, the consumer's requested services along with specified non-functionality held difficult due to huge amount of information available at broker's end. The proposed framework consisting of brokerage meta-model is useful to manage the inherent complex relations among the cloud providers in a much easier way. This is an efficient way to handle information with intelligence, so that the time of search is decreased.

After the inclusion of our proposal within the framework, it runs in the manner as depicted in Fig. 2. The query processing and extraction of service non-functional specification from the query is the preliminary need.

The following steps are depicted in Fig. 2 to illustrate the enhanced service discovery mechanism:

- First step considers the query from consumer.
- Functionalities are extracted from the query in second stage. Necessary SaaS, PaaS and IaaS are identified.
- Matching of necessary NFPs is done in this stage.
- The proposed meta-model is consulted for finding relevant feasible services.
- An NFP adjustment scheme may be needed for providing an integral solution.

The focus of the paper lies on the proposition of the meta-model and to describe the service discovery mechanism using the proposed work in Fig. 2.

3.1 The Proposed Meta-Model

The proposed meta-model grabs the non-functional information of the services and arranges those in a structured way. The meta-model is lattice based. This is immense helpful to use lattice as it contains all possible combinations of NFP in a structured way. It is important for better discovery. If a set of NFP at a specific level of lattice cannot satisfy the consumer-specified NFPs, it transfers to the next lower level or levels which are connected. These lower level nodes are nothing but the subsets of query NFPs. This actually increases the speed of discovery. Another benefit of lattice is that it maintains a hierarchy among the nodes.

The proposed structure contains all unique combinations of non-functional properties as elements. The elements together are defined as a set (say S). If the non-functional properties (NFP) are denoted by the letters (A, B and C), then

$$S = \{A, B, C, AB, AC, BC, ABC\}$$

Now each element in the set is actually a structure with varying dimension. The dimension of the structure depends on the number of NFPs, such as ABC is a three-dimensional structure as there are three NFPs (A, B, C). Node ABC has all services

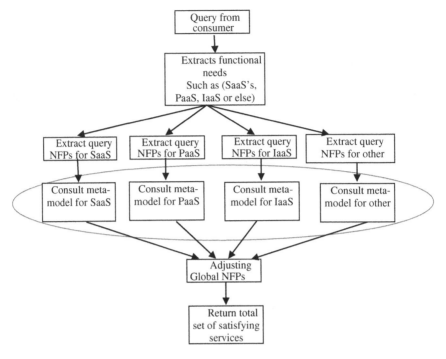

Fig. 2 Work flow of proposed cloud service discovery

whose specification has mentioned the values of NFP about *A*, *B* and *C*. Thus, the structure can be conceptualized by Fig. 3.

Two binary operations have been defined as "consolidation" and "diversification" on *S*.

Consolidation: It means accumulating common services from two different nodes into a single node (from *AB* and *AC* to *ABC*).

Diversification: It means a higher order node is decomposed into lower order nodes that is inclusion of more services in lower order nodes (*AB* to *A* and *B*).

Now from the definition of these two operations on the set S, following conclusions are drawn.

- Any node can be generated by higher level node using diversification or from lower node using consolidation. As the structure has unique nodes, reflexivity holds.
- Say nodes (*X* and *Y*) at level-*p* are generated by diversification from node *Z* at level-*p* + 1. Again applying consolidation on *X* or *Y* cannot regenerate *Z*. Antisymmetry holds.
- Say a consolidation on *M* at level-*p* generates *N* at level-(*p* + 1). Now the same is applied on *N* to generate *K* at level-(*p* + 2). It shows that *K* could be generated from *M* using consolidation. Hence, transitivity holds.

Fig. 3 Framework introduced

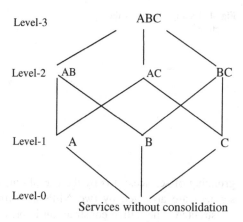

Level-3   ABC

Level-2   AB   AC   BC

Level-1   A   B   C

Level-0
Services without consolidation

S satisfies the reflexivity, anti-symmetry and transitivity properties. Thus, it is a poset. Any two combination of level-i in the structure can be mapped to a unique node in the level-$(i-1)$ which has all the common services of two selected level-i nodes, but the reverse is not true. Two nodes from level-i and node in level-$(i-1)$ are related by consolidation. So there exists unique least upper bound of any two nodes in the structure.

Similarly, any two combination of level-i is connected with a unique level-$(i+1)$ node by diversification, but also the reverse is not true. So there exists unique greatest lower bound for any two elements (nodes). The poset contains the unique least upper bound and unique greatest lower bound with respect to these two operations defined. Thus, it is proved as a lattice. Lattice will contain 2^N nodes in case of N NFPs. $(N+1)$ will be the number of levels in the structure.

The direction of consolidation and diversification within the lattice is also decided that helps into take dynamic decision at the time of discovery. For less NFP information and more number of services, **diversification** is indicated, i.e. the search will proceed from upper to lower nodes. Alternately, with more NFP information and lesser services, search will have lower to upper direction execution by the operation **consolidation**. This model works also for partial matching of NFP information though the existing ignores services with partial matches.

3.2 Description of the Framework

Consumer's query is defined here as q(FS, NFS), where FS is the functional specification and NFS is the non-functional specification. Again NFS can look like $(A = ``a_1", B = ``b_2")$ or $(A = ``a_1 - a_k", B = ``b_2 - b_j")$ where A and B are two NFPs. Number of NFPs are essentially 5, but can also vary within 8 to 10 for particular domain-specific SaaSs or general PaaS, IaaS, etc. The lattice in Fig. 3 has all services access pointers in the lowest level. Next level contains the one NFP

Fig. 4 Example node of the structure

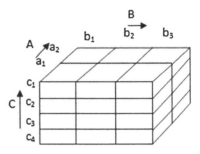

grouping of services; it is by the use of consolidation operation. In the next level, services are arranged by two NFPs at a time. This way the lattice structure is constructed. The lattice construction is primarily done at design time so the total overhead of the structure is only due to maintenance. Each of the nodes is a structure having dimensions. A typical structure is defined in Fig. 4.

The node *ABC* in Fig. 2 is internally looking like the above figure. Here, *A*, *B* and *C* are assumed as NFPs (such as cost, time and security). a_1 and a_2 are the different values assigned to the NFP *A*. The structure contains 24 cells ($2 \times 3 \times 4$). Each cell contains service access pointer through which the original services can be invoked. The very first cell in the above figure has edge labelled as c_1, a_1 and b_1 that means the cell contains the access pointers of the services whose description contains NFPs *A*, *B* and *C* and that too in the form of ($A = a_1$, $B = b_1$ and $C = c_1$).

This lattice model is introduced for containing service information in a structured way. Broker must have many lattices such as SaaS-lattice, PaaS-lattice and IaaS-lattice. A SaaS-lattice contains information regarding SaaS, which are accumulated and similar kind of structures formed for each of PaaS, IaaS, etc.

3.3 Service Discovery Algorithm

In this subsection, service discovery algorithm is defined. The described algorithm works within a lattice model. This is a generic algorithm that will work for every lattice within the framework. The following points are some prerequisite for understanding the algorithm

- All the NFPs are declared in the power of 2, i.e. for 3 NFPs *A*, *B* and *C*. So, $A = 2^0$, $B = 2^1$, $C = 2^2$ AB $= 2^0 + 2^1 = 3$. So all NFP combinations are declared as unique sums
- The algorithm works for all levels of the lattice.
- It is clear from the model that the number of NFP is equals the level of the lattice from where search is to proceed.
- The algorithm passes execution to next lower level if no services in *i*th level are matched with NFP specified.

- The algorithm will be recursively called several times until the services are found satisfying NFP.
- If failing to find all NFP satisfying services, the structure returns all possible combinations. For example, if the query specifies A, B and C and if at level-3 node ABC (in Fig. 3) has no matching services, then it will return services from AB, BC, AC, A, B, C nodes and the last node at bottom also.

Algorithm for searching services at the ith level of the lattice

Step 0:   [Initialize] Define unique weights to each NFPs. Set Source = level who called level-i, N = number of non-functional criteria passed by the user, Count = 1, Found[N] = false, fail = 1 (no of failed combination)

Step 1:   For the ith non-functional criteria do

Step 1.1:   If the user given non-functional criteria is valid then
                  Mark-no = i, [Mark it with i]
               End If

Step 2:   For all non-functional criteria, repeat step 1

Step 3:   Calculate unique sums for valid NFPs using defined weights.

Step 4:   Collect services with corresponding unique sums

Step 5:   For all existing services do
           Begin
                  If the all i non-functional criteria and functional criteria are simultaneously satisfied with user query, then
                  set a flag Found[i] = true
                  Print "The service found with id".
                  Endif
           End For

Step 6:   Increment count by 1.
           Repeat the steps from step 1 to step 4 for all combinations of valid ith NFPs

Step 7:   If Found[i] = true for all i then
                  If Source = manager then
                        Send a finish message to the manager.
                  Else
                        Send a found message to the Source.
                  End If
               Else
               If Source = manager then
               Call its previous level ($i − 1$).
               Else
               Send a not found message to the Source.
               End If
           End If

Step 8:   After all combination checked
                  If fail = no of all combination of all i NFPs then
                        Call ($i − 2$) level
                  End if

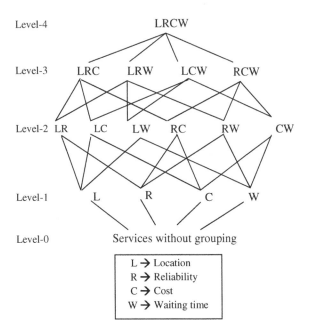

Fig. 5 SaaS-lattice for health care services

This multi-dimensional structure consists of several dimensions. These dimensions often contain hierarchy, such as location. If the service declaration has NFP, Location ="West Bengal" . Then, the hierarchy information is held in the lattice that "Kolkata" is in "West Bengal". So the services relevant to West Bengal will be retrieved, if the query demands it from Kolkata.

4 Illustration with Example

In this example, it is assumed that the framework consisting of five lattices for SaaS, PaaS, IaaS, DaaS (data as a service) and Communication as a service (CaaS). The example primarily is built upon a typical domain of SaaS. It is fixed as health care. So SaaS may include cost, waiting time, reliability and location as relevant NFPs with respect to health care services; particularly, emergency accidental care services are considered as the functionality. Thus, a SaaS-lattice formed containing emergency health services. Functionality of services includes diagnostic test services, general medicine, orthopaedics and nursing care. Figure 5 illustrates the SaaS-lattice.

Four NFPs are specified in the lattice of different services irrespective of functionalities, considering NFPs. A typical consumer query is of the following form:

Query: Brain scan imaging with maximum 2-h waiting time in Kolkata within cost Rs. 2,000 and the communication service with minimum delay to the services.

Solution: Here, a SaaS-lattice and another CaaS-lattice have to be exploited for satisfying the query. SaaS-lattice will be searched at level-2 and node CW. After finding a set of relevant services with respect to cost and waiting time from SaaS-lattice, a set of CaaS are to be found. Further, the CaaS-lattice is exploited with minimum delay that will connect any of the scan services satisfying the user NFP specification and Delay in CaaS.

5 Conclusion

Several discovery methods already exist in cloud domain; some of them are non-functional property-based mechanisms. In this approach, a methodology is proposed that works first to strike out the non-relevant NFP information. Further based on the remaining NFPs, any of the existing search algorithms can work.

Lattice that is a mathematical structure is used in cloud service domain. Few operations are there to explore the properties of the structure. Identification of the lattice framework, construction of it and searching on this model are the major contributions of this work. This model considers incomplete NFP information also and gives all possible combinations of those NFPs. These enable consumers to compromise at one of the NFPs and then subscribe the service where in existing works no such provision was there. Offered solution should be considered as a relevant one.

Scalability is another issue in this context. Huge number of services may occur within service cells in each structure, if there are no functional differences considered. In the proposed work, the domain of deployment is limited to a small set of cloud providers and they are related in some way. So the functionalities of services vary within a small set. However, the proposed work can be combined with any of the existing functionality-based solution of services discovery. As the number of NFPs within closed domains varying over 8–10, the construction and maintenance is easier.

In future, lattice of service units could be degenerated to work with only the important service NFP combinations and ignoring the rest. Besides, the multi-dimensional structure, referred here as node, could be exploited further. This may lead to the discovery of new set of operations and further optimizations. The identification of hierarchy and accordingly formation of new service at different granular level is an important research area where different mathematical models could be analysed.

References

1. Sundareswaran, S., Squicciarini, A., Lin, D.: A Brokerage-Based Approach for Cloud Service Selection. In: IEEE 5th International Conference on Cloud Computing, Honolulu, HI. ISBN: 978-1-4673-2892-0, June 2012
2. Subashini, S., Kavitha, V.: A survey on security issues in service delivery models of cloud computing. J. Netw. Comput. Appl. **34**(1), 1–11 (2011)
3. Lheureux, B.: Cloud Services Brokerages: the Dawn Of The Next Intermediation Age. Published in Gartner Blog Network, 8 Nov 2010
4. Lawler, C.M.: Cloud Service Broker Model, Green IT Cloud Summit, Washington, D.C, April 18, Sheraton Premier, Tysons Corner, http://www.greenitsummit.org/summit/ (2012)
5. Houidi, I., Mechtri, M., Louati, W., Zeglache, D.: Cloud service delivery across multiple cloud platforms. IEEE International Conference on Service Computing (SCC) (2011)
6. Nguyen, D.K., Lelli, F., Taher, Y., Parkin, M., Papazoglou, M.P., van den Heuvel, W.J. : Blueprint Template Support for Engineering Cloud-Based Services, Springer LNCS, Towards a Service-Based Internet. In: Proceedings of 4th European Conference, ServiceWave 2011 Poznan, Poland, 26–28 Oct 2011
7. Siebenhaar, M., Lampe, U., Lehrig, T., Žoller, S., Schulte, S., Steinmetz, R.: Complex Service Provisioning in Collaborative Cloud Markets, Springer LNCS, Towards a Service-Based Internet. In: Proceedings of 4th European Conference, ServiceWave 2011 Poznan, Poland, 26–28 Oct 2011
8. Tsai, W.T., Sun, X., Balasooriya, J.: Service-Oriented Cloud Computing Architecture. In: IEEE Seventh International Conference on Information Technology 2010
9. Mondal, A., Yadav, K., Madria, S.: Ecobroker: An Economic Incentive-Based Brokerage Model for Efficiently Handling Multiple-Item Queries to Improve Data Availability via Replication in Mobile-P2P Networks. In: 6th International Workshop in Databases in Networked Information Systems, pp. 274–283 (2010)
10. Buyya, R., Yeo, C., Venugopal, S., Broberg, J., Brandic, I.: Cloud computing and emerging it platforms: vision, hype, and reality for delivering computing as the 5th utility. J. Future Gener. Comput. Syst. **25**(6), 599–616 (2009)
11. World Wide Web consortium (W3C): Web Service Activity Statement. Retrieved from http://www.w3.org/2002/ws/Activity on 03 June 2007
12. Garg, S.K., Versteeg, S., Buyya, R.: A framework for ranking of cloud computing services. J. Future Gener. Comput. Syst. **29**(4), 1012–1023 (2013)
13. Qu, L., Wang, Y., Orgun, M.A.: Cloud Service Selection Based on the Aggregation of User Feedback and Quantitative Performance Assessment. In: IEEE 10th International Conference on Services Computing 978-0-7695-5026-8/13
14. Villegas, D., Bobroff, N., Rodero, I., Delgado, J., Liu, Y., Devarakonda, A., Fong, L., Masoud Sadjadi, S., Parashar, M.: Cloud federation in a layered service model. J. Comput. Syst. Sci. **78**, 1330–1344 (2012)
15. Heng, D.Y., et al.: A user centric service-oriented modeling approach. J. World Wide Web **14**(4), 431–459 (2011)
16. Narayanan, D., Flinn, J., Satyanarayanan, M.: Using history to improve mobile application adaptation. In: Proceedings of Third IEEE Workshop on Mobile Computing Systems and Applications, 2000
17. Balan, R., Satyanarayanan, M., Park, S., Okoshi, T.: Tactics-Based Remote Execution for Mobile Computing. In: Proceedings of the 1st International Conference on Mobile Systems, Applications and Services, ACM, pp. 273–286 (2003)
18. Kusic, D., Kandasamy, N.: Risk-aware limited look ahead control for dynamic resource provisioning in enterprise computing systems. In: Proceedings of the IEEE International Conference on Autonomic Computing, vol. 10(3); pp. 337–350 (2010)

19. Kofler, K., Haq, I.U., Schikuta, E.: User-Centric, Heuristic Optimization of Service Composition in Clouds. In: 16th International Euro-Par Conference, vol. 6271, Springer, Berlin, pp. 405–417 (2010)
20. Moens, H., Truyen, E., Walraven, S., Joosen, W., Dhoedt, B., De Turck, F.: Cost-effective feature placement of customizable multi-tenant applications in the cloud. J. Netw Syst. Manag. (2013). doi:10.1007/s10922-013-9265-5

A DWT-based Digital Watermarking Scheme for Image Tamper Detection, Localization, and Restoration

Sukalyan Som, Sarbani Palit, Kashinath Dey, Dipabali Sarkar, Jayeeta Sarkar and Kheyali Sarkar

Abstract The provision of image tamper detection, localization and restoration forms an important requirement for modern multimedia and communication systems. A discrete wavelet transform (DWT)-based watermarking scheme for this purpose is proposed in this communication. In our scheme, the original image is first partitioned into blocks of size 2 × 2 in which a 1D DWT is applied to produce a watermark which is embedded in four disjoint partitions of the image to enhance the chance of restoration of the image from different cropping attack-based tampers. The validity and superiority of the proposed scheme is verified through extensive simulations using different images of two extensively used image databases.

Keywords Discrete wavelet transform (DWT) · Least significant bits (LSBs) · Peak signal-to-noise ratio (PSNR) · Mean squared error (MSE) · Structural SIMilarity (SSIM) index

S. Som (✉) · J. Sarkar · K. Sarkar
Department of Computer Science, Barrackpore Rastraguru Surendranath College,
Barrackpore, West Bengal, India
e-mail: sukalyan.s@gmail.com

J. Sarkar
e-mail: sarkar.jayeeta9@gmail.com

K. Sarkar
e-mail: chelsea.kheyali9@gmail.com

S. Palit · D. Sarkar
CVPR Unit, Indian Statistical Institute, Kolkata, West Bengal, India
e-mail: sarbanip@isical.ac.in

D. Sarkar
e-mail: mampisarkar333@gmail.com

K. Dey
Department of Computer Science and Engineering, University of Calcutta,
92, APC Road, Kolkata 700009, West Bengal, India
e-mail: kndey55@gmail.com

© Springer India 2015
R. Chaki et al. (eds.), *Applied Computation and Security Systems*, Advances in Intelligent
Systems and Computing 305, DOI 10.1007/978-81-322-1988-0_2

17

1 Introduction

Tampering of digital media and its detection has been an interesting problem since long time. Its importance has increased with the stepping up of the use of digital media on the Internet. The volume of data transmission, especially that of images and videos, has gone up exponentially and has naturally drawn the interest of many including, unfortunately, fraudulent persons who would tamper with the transmitted data to suit their purpose. The detection of tampering followed by restoration of the original image is hence an important task. Most of the research carried out so far has been of tamper detection, while more recent work includes recovery of the image as well.

A number of digital watermarking schemes have been reported during the past decade for different purposes and considerations. In [1], an image tamper detection and recovery system has been developed based on the discrete wavelet transform (DWT) technique where some information has been extracted as the eigenvalue of the image and is embedded in the middle-frequency band of the frequency domain. Such embedding has been used for tamper detection and localization. In [2], a novel fragile watermarking scheme based on chaotic system for image authentication or tamper proofing is proposed. The watermark is generated by using pixel values as input values of a chaotic system, and a secret key controls a set of parameters of the chaotic system. A quantization function is introduced to embed and detect watermarks. This method can effectively detect minor alteration in a watermarked image. In [3], a tamper detection and retrieval scheme has been proposed. Special characteristic values of the low-frequency sub-band are embedded in the middle-frequency sub-bands. The embedded data with a digital signature and a public key are used to prove the authenticity of the image. Recovery with visually acceptable quality has also been achieved. In [4], the watermark of a particular image is generated from both frequency domain and spatial domain. The number of encoding stages of each DWT coefficient during the multistage encoding is taken as frequency watermark, and the mean values of blocks are stored as spatial watermark. The watermark is embedded into SPIHT encoded list of significant pixels (LSP) bit stream. By comparing the embedded watermark and the corresponding message extracted from decoded image, authentication is ensured. In [5], the semi-fragile watermark is designed from low-frequency band of wavelet-transformed image and is embedded into the high-frequency band by the human visual system (HVS). The robustness for mild modification such as JPEG compression and channel additive white Gaussian noise (AWGN) and fragility to malicious attack are analyzed. In [6], the proposed scheme extracts content-based image features from the approximation sub-band in the wavelet domain to generate two complementary watermarks. An edge-based watermark sequence is generated to detect any changes after manipulations. A content-based watermark is also generated to localize tampered regions. Both watermarks are embedded into the high-frequency wavelet domain to ensure the watermark invisibility. In [7], the original image is divided into two regions: region of interest (ROI), which is important region that requires protection against malicious modification, and region of embedding (ROE), which is the rest of the image where watermark sequence is embedded. In [8], dual visual watermarks using DWT and singular value decomposition (SVD) are presented. One is color image the same as original image, and the other is ownership watermark which is grayscale image. Both of them are embedded into original image using DWT-SVD to prove robustness. For recovery signal embedding, luminance signal and chrominance signal of original image were embedded into surplus chrominance space of original image using matrix transpose replacement embedding method. In [9, 10], two watermarks are used, generated from the low-frequency band and embedded into the high-frequency bands, one for detecting the intentional content modification and indicating the modified location and another for recovering the image. In [11], a multipurpose image watermarking method based on the wavelet transform is proposed for content authentication and recovery of the tampered regions where the original image is first divided into non-overlapping blocks and each block is transformed into the wavelet domain. The image features are subsequently extracted from the lowest frequency coefficients of each block as the first embedded watermark. Next, the whole image is decomposed into the two-level wavelet transform, and the orientation adjustment is calculated based on the wavelet coefficients in the middle-frequency sub-bands for image authentication. In addition, a logo watermark is embedded into the given middle-frequency sub-bands.

The rest of the paper is organized as follows. In Sect. 2, a brief introduction to DWT using Haar wavelet is given. In Sect. 3, the proposed scheme is presented wherein watermark generation, watermark embedding, and watermark extraction for the purpose of image tamper detection, localization, and recovery are explained. Section 4 demonstrates the experimental results with conclusions being drawn in Sect. 5.

2 Background

2.1 Discrete Wavelet Transform

The single-level 2D DWT decomposes an input image into four components, namely LL, LH, HL, and HH where the first letter corresponds to applying either a low-pass or a high-pass frequency operation to the rows and the second letter refers to the filter applied to the columns. The lowest frequency sub-band LL consists of the approximation coefficients of the original image. The remaining three frequency sub-bands consist of the detail parts and give the vertical high (LH), horizontal high (HL), and high (HH) frequencies. Figure 1 demonstrates single-level 2D DWT. For an one-level decomposition, the discrete 2D wavelet transform of the image function $f(x, y)$ can be written as follows:

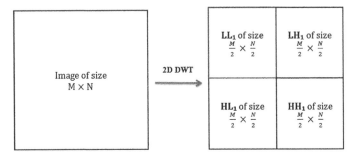

Fig. 1 Discrete wavelet transform

$$LL = [(f(x,y) \times \phi - x\phi - y)(2n, 2m)]_{(n,m) \in Z^2}$$

$$LH = [(f(x,y) \times \phi - x\psi - y)(2n, 2m)]_{(n,m) \in Z^2}$$

$$HL = [(f(x,y) \times \psi - x\phi - y)(2n, 2m)]_{(n,m) \in Z^2}$$

$$HH = [(f(x,y) \times \psi - x\psi - y)(2n, 2m)]_{(n,m) \in Z^2}$$

where $\phi(t)$ is a low-pass scaling function and $\psi(t)$ is the associated band-pass wavelet function. For computational simplicity, we have performed DWT using Haar wavelet.

3 Proposed Scheme

The proposed method has three distinct phases. Firstly, a watermark is generated from the image itself which is fragile to content modification as well as robust to common image processing after a preparation for doing so. Secondly, the generated watermark is embedded in the image. Finally, the watermark is extracted from the image (the one that has gone several degradations due to cropping attacks and/or noise attacks) to detect and localize tamper and recover the image as close as possible to the original one.

3.1 Watermark Preparation

A block mapping sequence is used to scramble watermark information. A 1D transformation algorithm, found in [12], shown in Eq. (1) is used to obtain a one-to-one mapping sequence where $X, X'(\in [0, N-1])$ the block number, k(a prime and $\in Z - \{$factors of $N\})$ is a secret key, and $N(\in Z - \{0\})$ is the total number of blocks in the image of size $N = 2^n \times 2^n$, $n \geq 2$, and $n \in N$.

$$X' = [f(x) = (k \times X) \bmod N] + 1 \tag{1}$$

A lookup table is constructed using the following algorithm to record the mapping address of each block in the image.

3.1.1 Block Mapping Address Generation Algorithm

1. Divide the image into non-overlapping blocks of 2×2 pixels.
2. Assign a unique nonnegative integer $X \in \{0, 1, 2, \ldots N - 1\}$ to each block from top left in row major order, $N = 2^{n-1} \times 2^{n-1}$.
3. Choose a prime number $k \in [1, N - 1]$.
4. For each block number X, obtain X' and its mapping block by Eq. (1). All the X's construct the lookup table.

A push-aside operation is used to modify the lookup table. The watermarks of the left half of the image are concentrated in the right half region of the image, and the watermarks of the right half of the image are concentrated in the left half region of the image. We simply push right the columns which originally belong to the left half and push left the columns which originally belong to the right half and thus result in a modified lookup table.

As an illustration, an image of size 8×8 is considered as the original image. The original image along with its corresponding block index matrix, lookup table generated using Eq. (1), and modified lookup table after push-aside operation is shown in Fig. 2.

3.2 Watermark Generation

Step 1: Decompose each 2×2 sized block by the DWT decomposition yielding from each block the approximation coefficient matrix LL_1 and the detail matrices HL_1, LH_1, and HH_1.

Step 2: The watermark is generated from the coefficient of the LL_1 sub-band of each decomposed block. As LL_1 wavelet coefficients may be beyond the recovery scope, its value must be adjusted. Therefore, the coefficients, after computation, are modified subsequently such that its value falls within the recovery range, as done in [5].

Step 3: The original image is divided horizontally and vertically into four equal parts. Let blocks A, B, C, and D be located at those four parts, respectively, such that C is situated at the opposite angle of A and D is situated at the opposite angle of B. Partner blocks of part A are located at the same position of part C and vice versa. Partner blocks of part B are located at the same position of part D and vice versa.

Step 4: The representative information of block A is constructed by extracting the five most significant bits (MSBs) of LL_1 sub-band coefficient of block A

(a)

30	58	62	64	65	57	55	56
37	119	114	115	115	116	111	106
38	121	115	109	112	110	114	104
37	108	121	109	114	113	105	109
38	115	124	118	110	118	106	112
37	114	118	106	113	109	113	111
36	110	107	113	103	114	110	112
36	110	115	103	110	113	113	102

(b)

30	58	62	64	65	57	55	56
37	119	114	115	115	116	111	106
38	121	115	109	112	110	114	104
37	108	121	109	114	113	105	109
38	115	124	118	110	118	106	112
37	114	118	106	113	109	113	111
36	110	107	113	103	114	110	112
36	110	115	103	110	113	113	102

(c)

0	1	2	3
4	5	6	7
8	9	10	11
12	13	14	15

(d)

1	14	11	8
5	2	15	12
9	6	3	0
13	10	7	4

(e)

11	8	1	14
15	12	5	2
3	0	9	6
7	4	13	10

Fig. 2 **a** The original image matrix; **b** the original image matrix subdivided into 2 × 2 non-overlapping blocks; **c** the original block matrix; **d** the lookup table; and **e** the modified lookup table after push-aside operation

and is then combined with (1) the representative information of block C and (2) the in-block parity-check bits and its complementary bit p and v, respectively, to construct the joint 12-bit watermark for blocks A and C. Similarly, the representative information of block B is used to construct the joint 12-bit watermark for blocks B and D.

The watermark generation technique is illustrated in Figs. 3 and 4.

3.3 Watermark Embedding

Two mapping blocks are needed to embed the joint 12-bit watermark of block A (or B) and its partner blocks C (or D). The lookup table helps find these mapping blocks. The watermark is embedded into the three LSBs of each pixel of a block. Suppose blocks $\overline{A}$ and $\overline{C}$ (or $\overline{B}$ and $\overline{D}$) are the two mapping blocks which are going to be used to embed the 12-bit watermark resulted from blocks A and C (or B and D). Both blocks $\overline{A}$ and $\overline{C}$ contain the same 12-bit watermark and the same embedding sequence in the corresponding locations. That is to say, for each block of size 2 × 2 pixels in the image, we have two copies of its representative information hidden somewhere in the image. Therefore, if one copy is tampered by any chance, we have two chances to recover this block from the other copy.

Figures 5 and 6 demonstrate the watermark embedding technique.

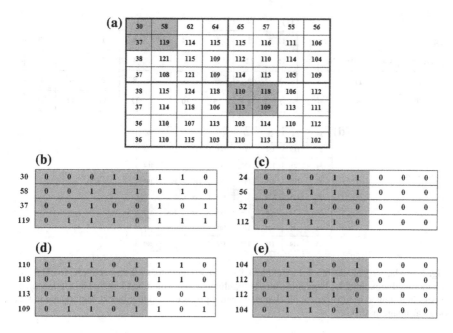

Fig. 3 **a** First two partner blocks (block 0 and block 10) in the original image matrix; **b** binary equivalent of each of the four pixels of block 0; **c** modified pixel values of block 0 after replacing three LSBs with 0s; **d** binary equivalent of each of the four pixels of block 10; and **e** modified pixel values of block 10 after replacing three LSBs with 0s

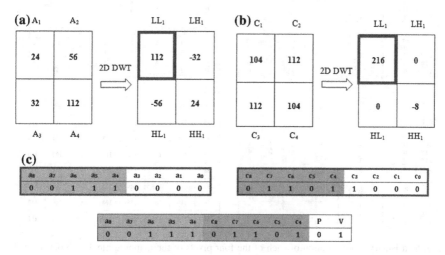

Fig. 4 **a** and **b** Application of 2D DWT using Haar wavelets into block 0 and block 10, respectively, resulting in the approximation coefficient matrix LL_1 and detail matrices LH_1, HL_1, and HH_1 and **c** the 12-bit watermark generated from the five MSBs of the LL_1 sub-band coefficient of block 0 and block 10 followed by a in-block parity-check bit P and its complement V

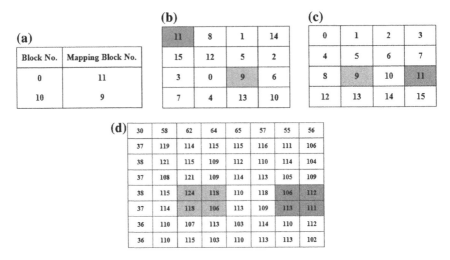

Fig. 5 **a** Mapping blocks block 11 and block 9 of block 0 and block 10, respectively, found from the modified lookup table; **b** mapping blocks highlighted in the modified lookup table; **c** mapping blocks highlighted in the original block matrix; and **d** pixels of mapping blocks highlighted in the original image matrix

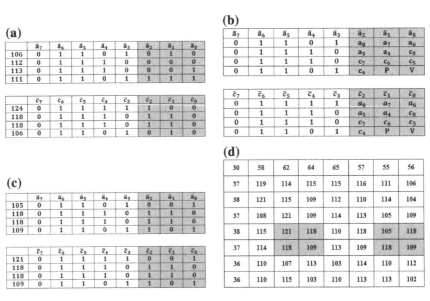

Fig. 6 **a** Binary representation of each of the four pixels of the mapping blocks—block 11 and block 9; **b** embedding of the same 12-bit watermark into block 11 and block 9; **c** modified block 11 and block 9 after watermark embedding; and **d** modified block 11 and block 9 in the original image matrix

3.4 Watermark Extraction: Tamper Detection, Localization, and Restoration

The watermarked image is tampered with different cropping attacks and covering and replacement attacks. Figure 7 represents the watermarked image of Fig. 6e cropped 25 % from center.

Tamper detection and localization A three-level hierarchical tamper detection and localization algorithm has been employed as proposed in [12].

Level 1 detection: For each non-overlapping block B of size 2 × 2,

1. Retrieve the 12-bit watermark information from the block.
2. Get the parity-check bits p and v, respectively, from the 11th and 12th bits of the retrieved watermark.
3. Perform exclusive-OR operation on the 10 MSBs of the 12-bit watermark, denoted by p'.
4. If $p = p'$ and $p \neq v$, mark block B valid; otherwise, mark it invalid.

Figure 8 demonstrates the level 1 tamper detection method.

Level 2 detection: For each block B marked valid after level 1 detection, check four triples (N, NE, E), (E, SE, S), (S, SW, W), and (W, NW, N) of the 3 × 3 neighborhood of block B. If at least one triple has all of its blocks marked invalid, mark block B invalid.

Level 3 detection: For each block B marked valid after level 2 detection, if at least five of the 3 × 3 neighboring blocks of block B are marked invalid, mark block B invalid.

Recovery of invalid blocks After the tamper detection process, all blocks in the image are marked either valid or invalid. Those invalid blocks need only to be recovered. A two-stage recovery scheme is applied for tamper recovery as follows:

Stage 1 recovery: For each non-overlapping block B of size 2 × 2 pixels which is marked invalid,

1. Find the mapping block of B from the lookup table, denoted by $\overline{B}$

(a)

26	58	58	68	67	58	50	60
39	114	116	117	118	117	108	109
35	122	115	106	114	108	115	106
38	109	124	0	0	118	108	105
34	116	121	0	0	114	105	118
38	118	118	109	118	109	118	109
35	106	106	114	98	114	106	114
38	109	118	102	111	114	118	102

(b)

26	58	58	68	67	58	50	60
39	114	116	117	118	117	108	109
35	122	115	106	114	108	115	106
38	109	124	0	0	118	108	105
34	116	121	0	0	114	105	118
38	118	118	109	118	109	118	109
35	106	106	114	98	114	106	114
38	109	118	102	111	114	118	102

Fig. 7 a Tampered image after cropping 25 % from the center of the watermarked image and **b** image in (**a**) with blocks highlighted

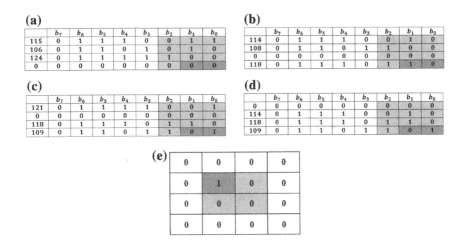

Fig. 8 Level 1 tamper detection and localization: **a** four pixels of block 5 with their binary equivalents; **b** four pixels of block 6 with their binary equivalents; **c** four pixels of block 9 with their binary equivalents; **d** four pixels of block 10 with their binary equivalents; and **e** localization of tampered block(s) after level 1 detection

2. If $\overline{B}$ is valid, then $\overline{B}$ is the candidate block, go to 5.
3. Find the mapping block of B's partner block, denoted by $\overline{\overline{B}}$.
4. If $\overline{\overline{B}}$ is valid, then $\overline{\overline{B}}$ is the candidate block; otherwise stop, leave block B alone.
5. Retrieve the 12-bit watermark information from the candidate block.
6. If block B is located in the upper half of the image, the 5-bit representative information of block B starts from the first bit (the leftmost bit) of the 12-bit watermark; otherwise, it starts from the sixth bit.
7. Pad four 0s to the end of the 5-bit representative information to form a new 9-bit coefficient.
8. Perform the inverse DWT operation based on this coefficient as the approximation coefficient which generates a new block of size 2×2.
9. Replace block B with this new block and mark block B as valid.

The method for stage 1 recovery is shown in Fig. 9.

Stage 2 recovery: Recover the remaining invalid blocks after stage 1 recovery from the neighboring pixels surrounding them. Corresponding to a central block B being processed, the 3×3 neighboring blocks can be found as directional triples (N, NE, E), (E, SE, S), (S, SW, W), and (W, NW, N) where each of the neighboring blocks being denoted as N_1–N_8 from NW to W in a clockwise manner. After the two-stage recovery process, lost blocks are reconciled by interpolating pixel values.

Figure 10 presents the reconstructed image of Fig. 7 after stage 2 recovery.

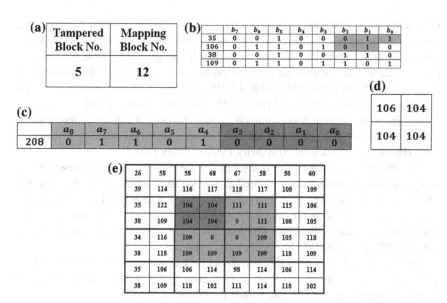

Fig. 9 Stage 1 recovery: **a** mapping block of the detected tampered block; **b** four pixels of the mapping block (block 12) with 5-bit information of block 5 embedded as watermark; **c** 5-bit information of block 5 padded with four 0s forming 9-bit approximation of block 5; **d** reconstructed block resulting from 2D inverse DWT on (**c**); and **e** recovered image after stage 1 recovery

(a)

26	58	58	68	67	58	50	60
39	114	116	117	118	117	108	109
35	122	106	104	111	111	115	106
38	109	104	104	111	111	108	105
34	116	109	109	109	109	105	118
38	118	109	109	109	109	118	109
35	106	106	114	98	114	106	114
38	109	118	102	111	114	118	102

(b)

26	58	58	68	67	58	50	60
39	114	116	117	118	117	108	109
35	122	106	104	111	111	115	106
38	109	104	104	111	111	108	105
34	116	109	109	109	109	105	118
38	118	109	109	109	109	118	109
35	106	106	114	98	114	106	114
38	109	118	102	111	114	118	102

Fig. 10 Stage 2 recovery: **a** and **b** the recovered image after reconciling the missing blocks by interpolating pixel values

4 Experimental Results

The performance and feasibility of the proposed scheme is examined through extensive tests carried out over USC-SIPI [13] and CSIQ [14] image databases which are collections of digitized images available and maintained by University of Southern California and School of Electrical and Computer Engineering,

Oklahoma State University, respectively. The images are chosen to prove the efficacy of the proposed scheme over various characteristics such as smooth areas, edges, textures, curvature, and regular and irregular geometric objects. The proposed scheme and the existing state of the art, considered for comparison, have been implemented using MATLAB 7.10.0.4 (R2010a) on a system running on Windows 7 (32 bit) with Intel Core i5 CPU and 4-GB DDR3 RAM.

The proposed scheme was examined against cropping attacks of different sizes. The performance of the proposed method is measured by the peak signal-to-noise ratio (PSNR) and Structural SIMilarity (SSIM) index [15].

The PSNR of a given image is the ratio of the mean square difference of two images to the maximum mean squared difference that can exist between any two images. It is expressed as a decibel value. An image with a PSNR value of 30 dB or more is widely accepted as an image of good quality. SSIM measures the similarity/dissimilarity between two images. For a watermarked image, greater value of PSNR and SSIM close to unity is expected.

Let $I_1(i, j)$ and $I_2(i, j)$ be the gray level of the pixels at the ith row and jth column of two images of size $H \times W$, respectively. The MSE between these two images is defined in Eq. (2), and PSNR is defined in Eq. (3).

$$\text{MSE} = \frac{1}{H \cdot W} \sum_{i=0}^{H-1} \sum_{j=0}^{W-1} |I_1(i,j) - I_2(i,j)|^2 \tag{2}$$

$$\text{PSNR} = 20 * \log_{10} \left(\frac{255}{\text{sqrt(MSE)}} \right) \tag{3}$$

The **SSIM** index between two images I_1 and I_2 as described in [15] is computed using Eq. (4):

$$\text{SSIM}(I_1, I_2) = \frac{(2\mu_{I_1}\mu_{I_2} + C_1)(2\sigma_{I_1 I_2} + C_2)}{(\mu_{I_1}^2 + \mu_{I_2}^2 + C_1)(\sigma_{I_1}^2 + \sigma_{I_2}^2 + C_2)} \tag{4}$$

where μ, σ, and σ^2 denote average, variance, and covariance, respectively, and C_1 and C_2 are constants as described in detail in [15].

4.1 Imperceptibility of Watermark

Imperceptible watermarks are invisible to naked eyes. If the embedded watermark is imperceptible, human eye cannot discriminate between the original image and its watermarked version. In the proposed scheme, the imperceptibility of the watermark has been examined for a wide variety of images in terms of PSNR and SSIM. For the watermarked images, greater value of PSNR (well above 35) and SSIM close to unity justify the imperceptibility of the watermark. A sample image

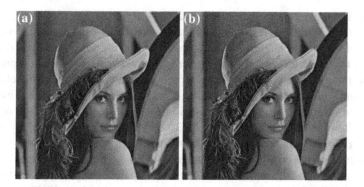

Fig. 11 a Original image of Lena; **b** watermarked image of (**a**)

Table 1 Comparison of PSNR and SSIM of watermarked images

Image name	Size	Ref. [12]		Proposed	
		PSNR (in dB)	SSIM	PSNR (in dB)	SSIM
Lena	512 × 512	41.44	0.93	41.44	0.93
Peppers	512 × 512	41.39	0.93	41.39	0.93
Baboon	512 × 512	41.30	0.98	41.31	0.98
Boat	512 × 512	41.35	0.95	41.32	0.95

of Lena and its watermarked version are shown in Fig. 11 where difference between the two images is hardly visible. In Table 1, the PSNR and SSIM between the original images and their watermarked versions using the proposed algorithm and the algorithm proposed by Lee and Lin [12] are presented.

4.2 Payload

The payload represents the size of the watermark that can be hidden in the image in terms of the number of bits per pixel (bpp). In our proposed algorithm, the size of the watermark is a function of the image size and block size. Here, the block size is of 2×2. For each block, a 12-bit watermark is embedded. For an image of size $H \times W$, the total size of the watermark embedded in the image is $\frac{H \times W}{2 \times 2} \times 12$ bits with a payload of $\frac{12}{2 \times 2} = 3$ bpp.

4.3 Performance Against Tampering

To evaluate the effectiveness of the proposed scheme against tampering, localize the tampered regions, and restore them back as close as possible to the original, the watermarked images were made to go through different types of tampers, viz. (1) *Direct Cropping* which can be classified into two sub-categories: (a) *cropping as a whole* where a single chunk is cropped from the image and (b) *multiple cropping* that includes *spread distribute* cropping where the cropping is spread all over the image and *chunk distribute* cropping where small number of relatively large chunks are cropped from the image; (2) *Object Insertion* where external objects are inserted into the watermarked image, and the object may be of large size, medium size, or small size; and (3) *Object Manipulation* where specific objects in the watermarked image are removed, destroyed, or changed.

Results of direct cropping (a) *Cropping as a whole*: Fig. 12 represents original image Lena of size 512×512, its watermarked version, different percentages of cropping attacks from center, and recovered images with their PSNR and SSIM values. From the result, we can see that the image can be restored up to a relatively good quality for cropping up to 60 %.

(b) *Multiple cropping*: Performance of the proposed scheme is evaluated against four different types of spread distribute tampering and eight different chunk distribute tampering. A total of 50 % of Peppers image is cropped. The cropped images along with corresponding tamper-localized and recovered images are shown in Fig. 13. Figure $13a_0$–d_0 represents spread distribute tampering, while chunk distribute tampering is represented in Fig. $13e_0$–l_0 for grayscale image of Peppers. The corresponding recovered images are presented in Fig. $13a_1$–l_1 along with their PSNR values. For brevity, the same test image Peppers, as in [12], is taken into consideration so that conclusions can be drawn that for different tamper distributions too, our proposed scheme outperforms the one in [12].

Results of object insertion One of the most common image tamperings by inserting objects is by copying/cutting regions of the watermarked image and pasting them into somewhere else in that image. The proposed watermarking system detects, localizes, and recovers the tampered regions of the images tampered by inserting small-, medium-, and large-sized objects as depicted in Fig. 14.

Results of object manipulation The watermarked image is attacked to remove, destroy, or change specific regions or objects in it. Figure 15 demonstrates three such attacks. The watermarked images are shown in Fig. 15a–c, the tampered images are shown in Fig. $15a_0$–c_0, the tamper-localized images are shown in Fig. $15a_1$–c_1, and the corresponding recovered images are shown in Fig. $15a_2$–c_2.

4.4 Comparative Study

To examine the advantages of the proposed scheme over the existing techniques, a comparative study is presented in this section. As we employed a block-based spatial domain watermarking scheme, a well-known work in this field proposed by Lee and Lin [12] is taken into considerations for performance comparison. In our approach, we have used the three LSBs of each pixel in the image for watermark embedding where the watermark has been generated from the LL_1 sub-band of

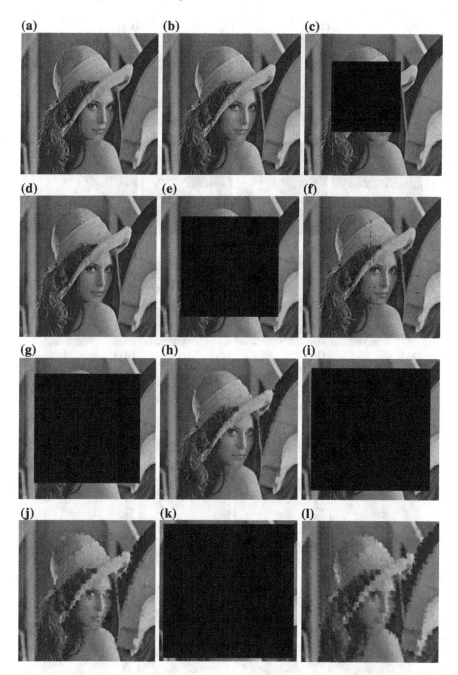

Fig. 12 a Original Lena image; **b** watermarked image of (**a**) with PSNR = 41.44 and SSIM = 0.93; **c** image in (**b**) tampered by 25 % cropping at center; **d** recovered image from (**c**) with PSNR = 35.51 and SSIM = 0.90; **e** image in (**b**) tampered by 50 % cropping at center; **f** recovered image from (**e**) with PSNR = 30.91 and SSIM = 0.85; **g** image in (**b**) tampered by 60 % cropping at center; **h** recovered image from (**g**) with PSNR = 30.07 and SSIM = 0.82; **i** image in (**b**) tampered by 75 % cropping at center; **j** recovered image from (**i**) with PSNR = 27.55 and SSIM = 0.7645; **k** image in (**b**) tampered by 90 % cropping at center and **l** recovered image from (**k**) with PSNR = 24.91 and SSIM = 0.67

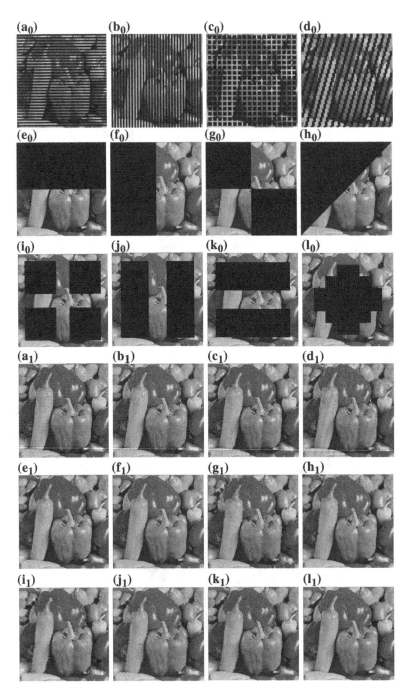

◀**Fig. 13** a_0–d_0 Spread distribute tampering, e_0–l_0 chunk distribute tampering of a total of 50 % in the watermarked image of Peppers (*grayscale*) of size 512×512, a_1 recovered image of (a_0) with PSNR = 32.19 dB, b_1 recovered image of (b_0) with PSNR = 30.58 dB, c_1 recovered image of (c_0) with PSNR = 33.12 dB, d_1 recovered image of (d_0) with PSNR = 28.76 dB, e_1 recovered image of (e_0) with PSNR = 32.89 dB, f_1 recovered image of (f_0) with PSNR = 33.30 dB, g_1 recovered image of (g_0) with PSNR = 27.56 dB, h_1 recovered image of (h_0) with PSNR = 30.19 dB, i_1 recovered image of (i_0) with PSNR = 29.30 dB, j_1 recovered image of (j_0) with PSNR = 29.95 dB, k_1 recovered image of (k_0) with PSNR = 31.39 dB, and l_1 recovered image of (l_0) with PSNR = 35.30 dB

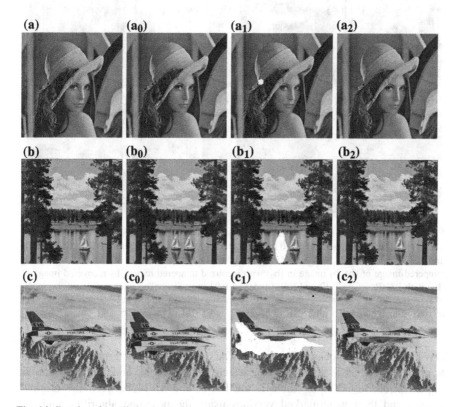

Fig. 14 Results of small-sized object insertion: **a** Watermarked image (*color*) of Lena of size 512×512, a_0 tampered image of (**a**) by inserting small flower on the hat, a_1 image in (a_0) with localized tampered region, and a_2 recovered image of (a_0) with PSNR = 41.07 dB and SSIM index = 0.94. Results of medium-sized object insertion: **b** Watermarked image (*color*) of sailboat on lake of size 512×512, b_0 tampered image of (**b**) by inserting a second sailboat on the lake, b_1 image in (b_0) with localized tampered region, and b_2 recovered image of (b_0) with PSNR = 39.61 dB and SSIM index = 0.0.95. Results of large-sized object insertion: **c** Watermarked image (*color*) of airplane of size 512×512, c_0 tampered image of (**c**) by inserting a second F-16 airplane, c_1 image in (c_0) with localized tampered region, and c_2 recovered image of (c_0) with PSNR = 33.92 dB and SSIM index = 0.90

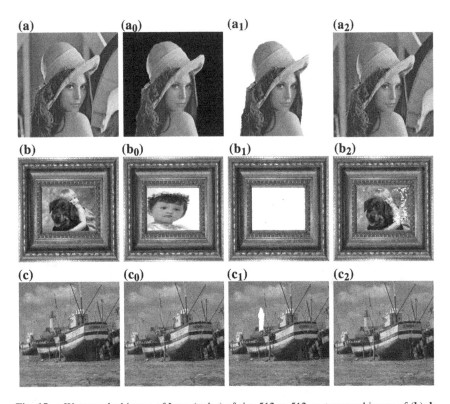

Fig. 15 **a** Watermarked image of Lena (*color*) of size 512 × 512, a_0 tampered image of (**b**), b_1 image of (b_0) with localized tampered region, a_2 recovered image of (a_0) with PSNR = 32.90 dB and SSIM index = 0.87, **b** a sample watermarked image (*grayscale*) of size 512 × 512, b_0 tampered image of (**b**), b_1 image in (b_0) with localized tampered region, b_2 recovered image of (b_0) with PSNR = 26.62 dB and SSIM index = 0.91, **c** watermarked image of boat (*grayscale*) of size 512 × 512, c_0 tampered image of (**c**), c_1 image of (c_0) with localized tampered region, c_2 recovered image of (c_0) with PSNR = 40.53 dB and SSIM index = 0.95

DWT transformed blocks of the image. The quality of our watermarked image in terms of PSNR is around 41.2 dB, which is acceptable, and the distortion is imperceptible to HVS. In Table 1, the PSNR and SSIM between the original images and their watermarked versions using the proposed algorithm and the algorithm proposed by Lee and Lin [12] are presented. Table 2 lists the comparison of the PSNR of the recovered image for the sample grayscale image of Lena for various tampered sizes and locations. When the tampered region is as small as 2.34 %, the performance of [12] is better than ours. But when the amount of tampered region (in percentage) grows gradually, it can be inferred from Table 2 that the proposed method performs better than the one in [12]. Table 3 presents the comparative study of the average PSNR values of images recovered from cropping attacks of different sizes for all the images available in the misc volume of USC-SIPI [13] image database (color images are converted to their grayscale versions).

Table 2 PSNR of recovered image relative to the tampered size and location (*test image* Lena)

Tamper (crop %)	Tamper location	PSNR (in dB)	
		In Ref. [12]	Proposed
2.34	Top	48.09	41.37
2.4	Center	39.48	41.05
8.01	Corner	41.42	41.13
9.7	Center	35.17	40.08
25.0	Left	33.45	40.44
34.0	Top	33.01	40.06
40.1	Center	27.97	33.53
50.0	Center	26.59	30.91
65.0	Center	24.57	29.21
70.0	Center	24.16	28.28
75.0	Center	23.43	27.55
80.0	Center	22.55	25.83
85.0	Center	21.28	25.50
90.0	Center	19.86	24.91
95.0	Center	18.05	20.96
97.0	Center	16.87	19.65
Average PSNR		28.50	31.90

Table 3 Comparative analysis of PNSR of recovered images from cropping attacks of different sizes

Crop (%)	PSNR (in dB)	
	Proposed method	Ref. [12]
10	37.57	33.80
20	35.74	31.84
30	34.72	29.91
40	33.54	29.40
50	31.40	27.05
60	29.72	25.86
70	28.15	25.10
80	26.91	22.35
90	23.76	19.47
Average	31.28	27.20

5 Conclusion

The simulation of various kinds of tampering with different images has demonstrated the superiority of the proposed method over that of the existing ones for different extents of tampering. The embedding of the DWT-based watermark in four regions of the image has been the major contribution of this work. Embedding in multiple regions has made the approach robust and helped it to perform well in even severe cases of tampering. Further research is being conducted to improve its performance for situations where very small areas are tampered.

References

1. Li, K.F., Chen, T.S., Wu, S.C.: Image tamper detection and recovery system based on discrete wavelet transformation. In: IEEE Pacific Rim Conference on Communications, Computers and Signal Processing, 26–28 Aug 2001. doi:10.1109/PACRIM.2001.953548 (2001)
2. Gang-chui, S., Mi-mi, Z.: Novel fragile authentication watermark based on chaotic system. In: International Symposium on Industrial Electronics, 4–7 May 2004. doi:10.1109/ISIE. 2004.1572034 (2004)
3. Chen, T.S., Chen, J., Chen, J.G.: Tamper detection and retrieval technique based on JPEG2000 with LL subband. In Proceedings of IEEE International Conference OD Networking, Sensing & Control, Taipei, Taiwan (2004)
4. Tsai, P., Hu, Y.C.: A watermarking-based authentication with malicious detection and recovery. In: 5th International Conference on Information, Communications and Signal Processing. doi:10.1109/ICICS.2005.1689172 (2005)
5. Tsai, M.J., Chien, C.C.: A wavelet-based semi-fragile watermarking with recovery mechanism. In: IEEE International Symposium on Circuits and Systems, ISCAS 2008. doi:10.1109/ISCAS.2008.4542097 (2008)
6. Qi, X., Xin, X., Chang, R.: Image authentication and tamper detection using two complementary watermarks. In: 16th IEEE International Conference on Image Processing (ICIP). doi:10.1109/ICIP.2009.5413681 (2009)
7. Cruz, C., Mendoza, J.A., Miyatake, M.N., Meana, H.P., Kurkoski, B.: Semi-fragile watermarking based image authentication with recovery capability. In: International Conference on Information Engineering and Computer Science. doi:10.1109/ICIECS.2009. 5363496 (2009)
8. Wang, N., Kim, C.W.: Tamper detection and self-recovery algorithm of color image based on robust embedding of dual visual watermarks using DWT-SVD. In: 9th International Symposium on Communications and Information Technology. doi:10.1109/ISCIT.2009. 5341268 (2009)
9. Yuping, H., Guangjun, G.: Watermarking-based authentication with recovery mechanism. In: 2nd International Workshop on Computer Science and Engineering. doi:10.1109/WCSE. 2009.856 (2009)
10. Hui, L., Yuping, H.: A wavelet-based watermarking scheme with authentication and recovery mechanism. In: International Conference on Electrical and Control Engineering (ICECE). doi:10.1109/iCECE.2010.86 (2010)
11. Wang, L.J., Syue, M.Y.: Image authentication and recovery using wavelet-based multipurpose watermarking. In: 10th International Joint Conference on Computer Science and Software Engineering (JCSSE). doi:10.1109/JCSSE.2013.6567315 (2013)

12. Lee, T., Lin, S.D.: Dual watermark for image tamper detection and recovery. Pattern Recogn. **41**, 3497–3506 (2008)
13. USC-SIPI image database: Available at http://sipi.usc.edu/database. Accessed on 1 Jan 2012
14. Computational Perception and Image Quality Lab, Oklahoma State University, www.vision. okstate.edu. Accessed on 1 Jan 2012
15. Wang, Z., Bovik, A.C., Sheikh, H.R., Simoncelli, E.P.: Image quality assessment: from error visibility to structural similarity. IEEE Trans. Image Process. **13**(4), 600–612 (2004)

Service Insurance: A New Approach in Cloud Brokerage

Adrija Bhattacharya and Sankhayan Choudhury

Abstract In cloud service domain, an acceptable standard of quality of service (QoS) must be maintained for subscribed services. The performance measurement of those cloud services is based on the satisfaction of customers with respect to the pre-defined QoS. Deviation of QoS as mentioned in SLA results dissatisfaction among users. A large numbers of business entities and consumers are involved in this service delivery process. In business environment, guaranteeing the QoS and establishing the service contracts are essential. However, for the service providers, it is challenging to maintain the QoS at run-time. Moreover, even if it is maintained, additional cost may be needed. Sometime a categorization among the consumers (premium or ordinary) is also required due to the limitation of the resources. Thus, the service management for ensuring the delivery with desired QoS at least for the premium consumers is necessary. This paper proposes a novel methodology termed as service insurances, which is incorporated into the service broker as a new module. The proposed concept is expected to ensure customer's satisfaction in context of a business application domain.

Keywords Service insurance · Cloud service life cycle · Risk modeling

1 Introduction

The increasing demand of cloud services at different associated quality of service (QoS) levels poses enormous challenges in service provisioning. In cloud architecture, broker plays an intermediate role of negotiating between cloud service

A. Bhattacharya (✉) · S. Choudhury
Department of Computer Science and Engineering, University of Calcutta, Kolkata, India
e-mail: adrija.bhattacharya@gmail.com

S. Choudhury
e-mail: sankhayan@gmail.com

© Springer India 2015

39

R. Chaki et al. (eds.), *Applied Computation and Security Systems*, Advances in Intelligent Systems and Computing 305, DOI 10.1007/978-81-322-1988-0_3

providers (CSP) and consumers. Typically in multi-cloud environment, broker is useful for executing some specific responsibilities [1] such as service level agreement (SLA) management, service provisioning, monitoring, and reporting. Figure 1 depicts a standard architecture of a service cloud broker. In spite of all necessary arrangement for ensuring service provisioning within a broker, the subscribed services may fail or the service with the assured QoS may not be delivered. The reasons behind these failures are mainly the over populated service demands, incompatibility among multiple service providers or lack of high-capacity infrastructures in smaller service provider's end [2], and inefficient load distribution. Customers use different hired services with various QoS levels in safety critical systems. Especially in those systems, QoS failure has high impact on customer applications, and as a result, this management issue demands higher attention.

The failure of service provisioning with requested QoS certainly hampers the reputation, and it may lead to a business loss of a corporate body. In a real-life situation, hundred percent availability of services with guaranteed QoS at all instances are practically impossible. A supporting mechanism will be necessary for the management of the service offering techniques such that the prioritized users (who are ready to pay premium) must be assured for getting the service with desired level of QoS.

In this work, we are offering the above-said solution, called "service insurance" through the broker. A proposed module is added within the existing framework of a broker to achieve this and is depicted in Fig. 1. The concept of paying insurance money for ensuring service (with specified QoS) leads to some new tasks such as premium amount calculation, categorization among users, and reserving resources for the prioritized users. The broker is supposed to guarantee the QoS to the consumers and helps providers also to decide which services are to be given more priority during maintenance or which customers are important with respect to

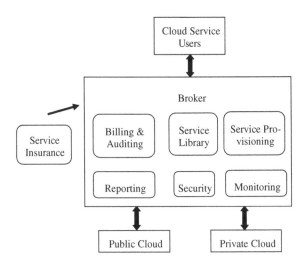

Fig. 1 Cloud service broker architecture

service offerings. Beside these, some of the other internal roles of the brokers are contract generation and management, failure handling, penalty decomposition, etc.

The broker works as the middleman between service providers and consumers during the negotiation. At the time of service publication, the insurance framework is initiated. The risk analysis and actuarial calculations are done, based on which the service insurance amount and periodical premiums are calculated. The negotiation between provider and consumer is being done in the presence of broker and is presented by a sequence diagram (Fig. 2).

In the negotiation phase, analyzing the consumer's query and QoS demand, total premium amount for all required services is calculated. If customer agrees on the amount, then contract document is finalized. Thus, as per the given contract, the providers take the responsibility for maintaining the QoS level and must utilize the insurance revenue for improving the provisioning. In contrary, the consumer may claim for penalties in case of failures. This is the main idea behind imposing the service insurances in addition to "pay per use" plan. Different rates can be fixed for multiple tariff options for insurances and they are directly related to the given QoS. Guaranteeing the higher assurances, insurance premium goes higher.

This presented work is concentrating on the risk analysis and modeling issues for cloud computing environment. The concept insurance is built up based on the user dissatisfaction, and in turn, the dissatisfaction is highly dependent on the risk analysis. Overall, description of the framework is given in Sect. 3. The reasons behind identification of the risks and service failures are discussed in Sect. 4. Section 5 proposes a model for risk calculation and the Sect. 6 concludes.

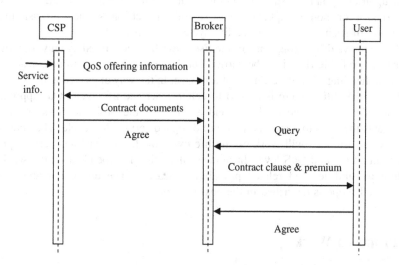

Fig. 2 Sequence diagram for cloud service insurance framework

2 Related Work

The proposed concept of insurance can be achieved after successful implementation of some prerequisite tasks. The risk identification, analysis, and modeling are some major issues on which the concept of insurance is build up. The risk modeling is our current focus of work, and thus, in this section, we have mentioned few notable works in this specific issue.

In [3], a set of threats such as natural disaster and cloud malware has been identified. But all the threats declared here are not directly related to cloud services. In [4], the risk mentioned as hardware failure is decomposed into poor I/O performance, poor CPU utilization, etc. In [5], a number of risks identified and discussed with respect to SaaS, PaaS, etc. Also, some risk terms are identified; based on these terms, each risk factor is described. These terms include probability of risk occurrence, impact of the risk, and categories of risk. Each of these three terms has three levels associated, namely high, medium, and low. Based on these terms, each risk is defined for SaaS, PaaS, and IaaS. However, no method for risk assessment and modeling is discussed in this approach. A set of resource management policies and related risks is discussed in [6, 7]. These are helpful for identifying risks, and the objective of the proposed policies is to make resource provisioning better. In [8], the internet jurisdiction issues have been discussed. The impact of geographical and jurisdictional information is described here. A few legal issues are also addressed. The business and organizational policy-related issues are discussed in [9]. The collaboration among service providers is important in service cloud federation. The collaboration among providers is highly dependent upon compliance issues. Violation of any contract clause, resulting loss of governance or non compliance, has severe impact on service provisioning. Inclusion of such issues in service life cycle is also needed.

In the above discussion, some of the risks have been identified. An investigation about the probable risks is to be carried out. Necessary risks are to be considered for apprehending the failure causes of a service. In the way of assessing the service failure probabilities, there is a need to define a service life cycle. Our approach should take care of the modeling and analysis of the probable risks and also associates the risks to a specific phase of designed service life cycle. The concept of service insurance will work on the life cycle model for ensuring service provisioning at assured QoS level. Insurance calculation is dependent on the service failure possibilities and related probability measure. The discussion about the mentioned scope is described in the next section.

3 Proposed Work

In this present work, the outline of the proposed concept, i.e., "service insurance" is discussed. The sub-tasks for achieving the goal are identified, and all these tasks are conceptualized within a module. This module is assumed to be added within

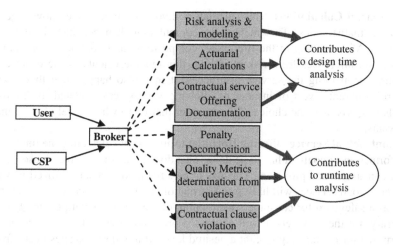

Fig. 3 Service insurance components within broker

broker architecture for offering the said solution. The main focus of this paper is to identify the risks and associate the risks with a specific phase of service life cycle. Moreover, a measurement mechanism to assess the vulnerability of a service in terms of failure (service provisioning without desired QoS) is proposed, and this will be used for actual premium amount calculation.

Each set of required cloud services needs a guaranteed QoS for proper delivery. It also depends on the QoS requirement of the consumer. Thus, for getting service with guaranteed QoS, consumer has to subscribe, i.e., insure services. In broker-based cloud service architecture, the broker should play the role of smooth service delivery as per the requirement. As per our proposal, a broker has to include six more modules for handling service insurance procedure. In Fig. 3, the boxes depict the newer modules to be included. Some of the components are designed as contributor at design time analysis. Rests are contributors to run-time analysis. In the next section, purpose of all these tasks is discussed.

3.1 Component Description of the Framework

Risk Analysis and Modeling: Risks involved for insuring a set of services is dependent on some assessment parameters. These parameters are very much application specific. Modeling these factors and calculating the impact of them on services performance are important. The idea of service life cycle is considered here. This starts with the publication of service and ends at the service feedback receipt. Risks are widely distributed over the different phases of life cycle. Risk can have two types of reasons. One set of cause is fixed in nature and the rests are variable (Random risk).

Actuarial Calculations: This calculation involves typically the knowledge of actuarial mathematics. This method of computation demands historic data on services. After looking at the life cycle, composition, and contract details along with some inferences from risk modeling, this module calculates the exact cost amount for availing the service insurances. It may also happen that there exist several different levels of insurances allowing varying costs related to different levels of QoSs. It is the choice of a consumer that up to which level of assurance he wants.

Contractual Service Offering Documentation: After calculating the final cost, the formation of contract has to be done. This document contains the clauses from both consumers and providers. This has to be structured in a standardized way so that any cloud services will have same format of contracts. The contract document acts as a safeguard for consumers mentioning offered services with given QoS. In contrary, for the providers, it will assure the profit ratios as well as the additional overheads for committing QoS at a desired level. If any discrepancies occur from provider's side, then there are provisions for penalties. But the key point is the consumer and provider must agree on the standardized rates and clauses via CSB.

Penalty Decomposition: The penalty decomposition is an existing idea. There are several existing methodologies for penalty decomposition among sub-providers. Any of the standard decomposition method can be adapted by the framework for actually assessing the failures and distributing those responsibilities. Further, the penalties are calculated and collected from multiple providers.

Quality Metrics Determination with Respect to Queries: The quality requirement of consumers (queries) varies often. Thus, for a specific query, the set of services and the corresponding level of QoS determination are important. It works on demand. The priority given by the users to a specific parameter (s) is also an important issue at metric determination phase.

Contract Clause Violation: Violation of any contract rules by both the provider and consumer results same illegal affair. This is the phase that would indicate that something out of contract has been occurred and necessary steps are to be taken further.

3.2 Workflow of the Framework

This section illustrates the working procedure of the service insurances as depicted in the Fig. 4. It has three entities: service provider, consumer, and broker. These components of broker are highlighted by the gray boxes. The operations and responsibility carried are labeled by the edges. For the ease of understanding, each of the six broker module mentioned in Sect. 3.1 is marked in dark colors. Edges 1–5 are executed at the design time. It is the conversational phase between broker and service providers. On the other hand, edges 6–9 are run-time execution phases and this is query specific also. This conversation is between consumer and broker.

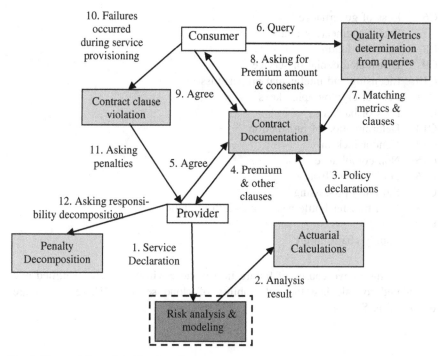

Fig. 4 The workflow of the framework

Edges 10–12 are executed after the failure occurrences. We have initiated the concept of service insurance and also provide the framework for accomplishing this. But the box marked with dotted line is the area of concentration in this paper.

4 Risk Analysis

In this paper, only the risk analysis and modeling component from Fig. 4 are discussed in detail. It is the first stage of analysis. The potential risks are needed to be identified with respect to each stage of the cloud service life cycle. Irrespective of service functionalities (IaaS, PaaS, SaaS, etc.), some risks are inherent within service life cycle. We consider the following definition of risk in this context. *Risk is failure of offered services for any cause or deviation from ensured level of QoS in service offering. These risks are identified as follows:

C1   Data unavailability
C2   Natural disaster
C3   Malicious insider (provider)
C4   Data leakage
C5   Cloud malware

C6   Loss of governance
C7   Social engineering attack
C8   Isolation failure
C9   Distributed denial of service
C10   Licensing and intellectual property issues
C11   Loss of cryptographic keys
C12   Loss of backups
C13   Determination of jurisdiction
C14   Vendor lock in
C15   Non-compliance
C16   Poor CPU utilization
C17   Slow I/O processing
C18   Data transfer bottle-neck bandwidth limitation
C19   Bugs
C20   Supply chain failure

All of the above causes of failure in a cloud environment are identified and considered to calculate the vulnerability of cloud services. These causes are described in Sect. 4.1.

4.1 Causes Behind Risk Identification

C1 may often cause failure. The unavailability of data [10] may cause a stop of an ongoing service execution. It may reduce the speed of computing services or make any service unauthorized and unavailable. Cause C2 can happen anywhere anytime, though this can be thought under the random risks. Sometimes, there exists some pseudo-provider or mal insider within the cloud federation. These providers can harm execution and QoS offering that may ultimately lead to service failure. Data leakage [11] is one of the fatal attacks that may happen to cloud data center, and this failure causes insecure transaction or mal functioning service offering. Another attack in this course is metadata spoofing attack [12]. It results the user's dissatisfaction. Cloud malwares [13] are often injected into the transmission data and SQL codes that results failure.

The business behind the cloud provisioning is running in relation to some profit–loss equation. Each provider has some business policy and a governing entity [14] for proper management. The lack of responsibility and sometimes ambiguity among business process declaration can cause service mal functioning. Exponential increase in number of users compelled the IT services to distribute data in cloud. This is a scope of exposure for an attacker who aims to confuse the users about popular services. In social engineering attack [15], attacker often confuses the employees of provider organization with bad intensions to expose their business services to fatal errors. In public cloud service domain, different users share infrastructures and they often feel lack of privacy with respect to data and confidential business information. Sharing policies [16] at data center and clouds plays important role here. But failure can also occur by this wing.

Distributed denial of service (DDOS) [17] is a type of attack that makes resources unavailable to tenants. There exist three types of licensing among cloud that are for user, device, and enterprise. Often, some provider has false licenses [18] and intends to harm user's information and security. Cloud service provider is always liabilities for any content upload to a cloud, and always, the provider has to take "notice and takedown" [19]. In browser security for cloud users, it is necessary to protect encryption keys, but loss of it is a serious threat toward security [12].

Loss of saved data from data center due to any reason is alarming. But this situation can be handled by some replication [20] policy along with network-distributed storage mechanisms. But deletion or loss of backup data puts service offerings in a critical situation. In determining crime over internet and its juris-diction [8] is really difficult. The geographic location and applicable legal theories often vary due to lack of standardization. However, risks due to this reason till date persist. Data transfer between cloud providers held difficult due to vendor lock in. It often solved by several mechanism, but it can be considered to contribute significant risks in service offering [21].

Another important part of cloud computing is compliance especially [9] reg-ulatory compliance. This issue can hamper inter-organizational collaborative offerings and discontinuation of services. I/O processing speed and CPU utilization [22] are two system performance parameters. Under performance of any of the two hampers computing and infrastructure services highly. Other service offerings are also get affected by these. On-demand applications and high bandwidth-consuming applications [23] often fail due to limitation of bandwidth and poor management of priority policies among users. There exist code bugs in any application. These are one of the reasons of service failure. BPaaS [24] is one of the popular services in cloud that is highly dependent on supply chain management. Any failure of supply chain [4] will affect this service offering. Beside that lack of control and visibility in any supply chain can cause risk at services.

4.2 Service Life Cycle and Risks

In Fig. 5, the cloud service life cycle is described, and the above-mentioned causes are associated with each stage of the life cycle. There are four compo-nents within the life cycle that are responsible for different cloud liabilities. These four are Service Governance, Service Development, Service Release and Communication, and Service offering. All the components and associated risks are described here.

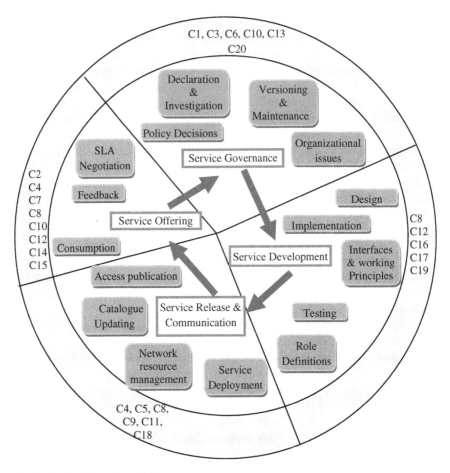

Fig. 5 Service life cycle and risks

4.3 Service Governance and Related Risks

Service Governance has service declaration and investigation as a component. It is responsible for service announcement and investigating related business issues. Different collaborative issues are handled within organizational issues. Business policy and company decisions are governed by the policy decision-making component. Including customer feedback and managing version-related information are also the responsibility of this phase. From the identified causes list in Sect. 3, C1, C3, C6, C10, C13, and C20 are the causes that may result failures in this phase.

4.4 Service Development and Risks

This phase has five specific responsibilities. Service design and role definitions are most important of the five. Implementation of services is associated with the risks related to the infrastructure and software-related failure causes. Designing and coding of interfaces has some of the hardware issues and some legal issues also. Testing is done for bug removal and other error corrections, but still there exist some causes that can result failure at this phase. These are C8, C12, C16, C17, and C19.

4.5 Service Release and Communication and Related Risks

This phase has four components. Service deployment and access publication in service catalog are two important responsibilities. Network resource management is another issue. This component basically handles communication in cloud service provisioning through bandwidth management. Network-related risks are all valid for this stage. Additionally, cryptographic issues and malware of clouds also result risk in this phase. C4, C5, C8, C9, C11, and C18 are valid causes in this phase.

4.6 Service Offering and Risks

Service offering phase has three responsibilities. SLA negotiation is an important one. In service level agreement (SLA), the service provider and consumer agree upon QoS levels. This is significant to judge the user satisfaction. Customer feedback also used to measure user satisfaction. This phase is prone to the attack caused by C2, C4, C7, C8, C10, C12, C14, and C15.

5 Risk Modeling

There exist 20 causes for which a service may fail. Now in this section, we are modelling the risk probability associated with the 20 different causes. In general, there exists recovery mechanism associated with a specific failure. So it is to be noted that a service actually fails due to a cause if the cause has been occurred and the recovery from the cause has been failed also.

Here, Y_i is a random variable that is denoted as follows:

$Y_i = 1(C_{ith}$cause in the list occurred)with probability p_i

$0(C_{ith}$ cause in the list didn't occur)probability $(1 - p_i)$

Where, $0 < p_i < 1$ and $i = 1, \ldots, 20$

i.e. $P(Y_i = 1) = p_i$

It is clear from the foundation that Y_i follows Bernoulli's distribution with parameter p_i. p_i indicates the probability of occurrence of the C_{ith} cause. It is preliminarily assumed that the causes occur independently. Thus, it is clear that Y_i is independently distributed. It is obvious to state that occurrences of any of the causes not always result service failure. There are some recovery mechanisms. The failure occurs when the recovery is also failed. So the probability of cause occurrence is different from that of service failure due to that cause. Let us consider r_i to be the probability of recovery from the C_{ith} cause.

$$P(\text{recovery from } C_{ith} \text{ cause}) = r_i$$
$$P(\text{not recovery from } C_{ith} \text{ cause}) = 1 - r_i$$

The probability of service failure from C_{ith} cause $= P(A_i \cap N_i)$
Where,

A_i the cause C_i occurs
N_i is an event that the failure from C_{ith} cause is not recovered

Then, by theorem of conditional probability, we get,

$$P(A_i \cap N_i) = P(N_i) \cdot P(A_i|N_i)$$
$$= (1 - r_i) * p_i$$

Let us consider another random variable X_i as Bernoulli variable which signifies that the value of X_i will be 1 if the failure is from C_{ith} cause. So,

$X_i = $ 1 with probability $(1 - r_i) * p_i$

0 with probability $1 - \{(1 - r_i) * p_i\}$

Again, we are declaring the following variable Z as

$$Z = \sum_{i=1}^{20} X_i$$

Basically, Z is sum of all failures. Z can take values as 0, 1, ... , 20. So probability of service failure from any cause is formulated by

$$P(Z > 0) = 1 - P(Z = 0)$$
$$= \prod_{i=1}^{20} (1 - r_i) * p_i$$

Initial probabilities r_i and p_i are to be estimated. p_i can be obtained by fitting Poisson model to the data obtained on failure causes. Each cause is fitted as Poisson variable as the failure occurrences are happened to be infinite in nature. After that normal approximation of failures will result the value of r_i.

Here, it is assumed that the cause occurrences are independent. Further study can be done on interdependence of causes. This analysis has to be done by the broker for each service. Each service will have different values of $P(Z > 0)$. This probability value is considered here as vulnerability quotient of that service. The vulnerability quotient is considered here to calculate the service availability at assured level of QoS. Further, the insurance calculation and sum of premium are calculated based on that quotient. So this risk modeling satisfies the primary need on which the insurance calculation and the calculation life expectancy of services depend.

6 Conclusion

This paper is to provide the outline of service insurance concept. The main focus of this paper is identification of risks and association of those in the service life cycle. Based on this, the detailed risk modeling and failure estimation are also discussed here.

There exist a few other components in the service insurance model that are yet to be explored. However, the components identified have high impact on business management. This will pioneer an avenue toward newer business systems. The deduction of several new business rules is in the future scope of this work. This risk model may be extended by mitigating failures from interdependent causes. Similarly, application of actuarial mathematics in the domain, a comparative study among several actuarial and risk modeling, and forecasting methods are the areas for future explore.

References

1. Lawler, C.M.: Cloud service broker model-sustainable governance for efficient cloud utilization. Green IT Cloud Summit, Washington, D.C, Sheraton Premier, Tysons Corner, 18 Apr 2012
2. Hassan, M.M., Song, B., Huh, E.N.: A market-oriented dynamic collaborative cloud services platform. Ann. Telecommun. 65, 669–688 (2010). doi:10.1007/s12243-010-0184-0

3. Cloud computing vulnerability incidents: a statistical overview. Cloud Vulnerabilities Working Group, Cloud Security Alliance (2013)
4. Pearson, S., Benameur, A.: Privacy, security and trust issues arising from cloud computing. IEEE Second International Conference on Cloud Computing Technology and Science (CloudCom). IEEE (2010)
5. Cloud computing-benefits, risks and recommendations for information security. European Network and Information Security Agency, December 2012
6. Gmach, D., Roliat, J., Cherkasovat, L., Belrose, G., Turicchi, T., Kemper; A.: An integrated approach to resource pool management: policies, efficiency and quality metrics. In: International Conference on Dependable Systems & Networks: An-chorage, IEEE, Alaska, 24–27 June 2008
7. Toosi, A.N., Calheiros, R.N., Thulasiram, R.K., Buyya, R.: Re-source provisioning policies to increase IaaS provider's profit in a federated cloud environment. In: IEEE International Conference on High Performance Computing and Communications 2011
8. Ward, B.T., Sipior, J.C.: The Internet jurisdiction risk of cloud computing. Inf. Syst. Manag. 27(4), 334–339 (2010)
9. Farrell, R.: Securing the cloud—governance, risk, and compliance issues reign supreme. Inf. Secur. J.: A Global Perspect. 19(6), 310–319 (2010)
10. Cidon, A., et al.: Copysets: reducing the frequency of data loss in cloud storage. Presented as part of the 2013 USENIX Annual Technical Conference. USENIX, (2013)
11. Wang, C., et al.: Privacy-preserving public auditing for data storage security in cloud computing. INFOCOM, Proceedings IEEE, (2010)
12. Jensen, M., et al.: On technical security issues in cloud computing. IEEE International Conference on Cloud Computing. CLOUD'09. IEEE (2009)
13. Jamil, D., Zaki, H.: Security issues in cloud computing and countermeasures. Int. J. Eng. Sci. Technol 3(4), 2672–2676 (2011)
14. Fortis, T.F., Munteanu, V.I., Negru, V.: Steps towards cloud governance. a survey. In: Proceedings of the ITI 2012 34th International Conference on Information Technology Interfaces (ITI), IEEE (2012)
15. Bezuidenhout, M., Mouton, F., Venter, H.S.: Social engineering attack detection model: SEADM. Information Security for South Africa (ISSA). IEEE (2010)
16. Raj, H., et al.: Resource management for isolation enhanced cloud services. In: Proceedings of the ACM workshop on Cloud computing security. ACM (2009)
17. Bakshi, A., Yogesh, B.: Securing cloud from ddos attacks using intrusion detection system in virtual machine. In: Second International Conference on Communication Software and Networks. ICCSN'10, IEEE (2010)
18. http://searchcloudcomputing.techtarget.com/feature/Cloud-computing-licensing-Buyer-beware
19. Cordell N.: Intellectual property in the cloud. Allen &Overy LLP, May 2013
20. Vrable, M., Savage, S., Voelker, G.M.: Cumulus: filesystem backup to the cloud. ACM Trans. Storage (TOS) 5(4), 14 (2009)
21. Marinos, A., Briscoe, G.: Community cloud computing. Cloud computing, pp. 472–484. Springer, Berlin (2009)
22. Evangelinos, C., Hill, C.: Cloud computing for parallel scientific HPC applications: feasibility of running coupled atmosphere-ocean climate models on amazon's EC2. Ratio 2(2.40), 2–34 (2008)
23. Niu, D., et al.: Quality-assured cloud bandwidth auto-scaling for video-on-demand applications. INFOCOM, IEEE (2012)
24. Crowe, H., Chan, W., Leung, H., Pili, H.: Enterprise risk management for cloud computing. Committee of Sponsoring Organizations of the Treadway Commission, June 2012. http://www.coso.org/documents/Cloud%20Computing%20Thought%20Paper.pdf

Using Semiformal and Formal Methods in Software Design: An Integrated Approach for Intelligent Learning Management System

Souvik Sengupta and Ranjan Dasgupta

Abstract The use of graphical methods such as unified modelling language (UML) in conjunction with formal methods such as Vienna development method (VDM) can be significantly beneficiary in the software design phase due to their complimentary features. UML diagrams are very useful in communication among different stakeholders, but at the same time, being semiformal in nature, they lack formal syntax and preciseness due to textual description in notations. This makes it challenging to verify the design against the requirements. Conversely, a formal specification language like VDM-SL has the advantage of preciseness an unambiguous modelling, but unable to provide ease of understanding like UML. This paper presents a methodology that integrates the use of UML and VDM-SL in software design phase and also proposes a verification technique for the design artefacts with the requirements. A case study of intelligent learning management system (ILMS) is used in this paper to illustrate the proposed work.

Keywords VDM-SL · Software design · UML · Design verification

1 Introduction

The software design artefacts demonstrate how to fulfil the requirements and guide the implementation of that software item. UML notations are widely accepted and used for visualizing models in different phases of developments, from abstraction of requirements to detail design of them [1, 2]. The objective of this work is to propose a design and specification methodology that integrates both an industrial

S. Sengupta (✉)
Bengal Institute of Technology, Kolkata, India
e-mail: mesouvik@hotmail.com

R. Dasgupta
National Institute of Technical Teachers Training and Research, Kolkata, India
e-mail: ranjandasgupta@ieee.org

© Springer India 2015   53
R. Chaki et al. (eds.), *Applied Computation and Security Systems*, Advances in Intelligent
Systems and Computing 305, DOI 10.1007/978-81-322-1988-0_4

standard such as UML and a formal description language such as VDM-SL, enhancing the relationship between the two methodologies. Integrating a semi-formal graphical modelling technique with a formal development method results in a development framework that supports rigorous analysis of the design model and also verification of the design artefact against the requirement specification. However, the use of formal methods in design phase entails formal methods to be used in requirement specifications. We presume that the requirement artefact contains use case and VDM-SL specification, and then, we design the system by elaborating the use case diagram into class diagram, activity diagram, sequence diagram and also VDM-SL specification. The verification model has two components: first, we check for consistency and continuity among these UML diagrams at design phase, and then, finally, they are verified against the requirements elicited and specified in the RE phase. Instead of mapping the design articles against requirements written in natural language (NL) statements, we use conceptual graph (CG) for this purpose. Since comparison of requirements written in NL statements for traceability is difficult to perform, CG can work as an intermediate stage, which can be easily compared with NL statements.

The work presented in this paper is arranged in the following manner. Section 2 details the related works and the scope of work in this domain. Section 3 describes the proposed framework. Next, Sect. 4 gives overview of a simple case study and is referred by the remaining parts of the paper. Section 5 states a requirement specification as we expect from the RE phase. Section 6 depicts different design artefacts, and finally, Sect. 7 illustrates the design verification technique.

2 Related Work

Integrating formal methods with semiformal or graphical methods in software development is not an novel idea, and some forms of combinations are available in a reasonable number of works [3]. However, as pointed out by several researchers, there are many reasons why they are still to come into practice. One of that is the lack of tools that support for integrating formal techniques with traditional semiformal methods. Another reason is the apprehension of difficulty in following complex mathematical expressions and to relate abstract descriptions with real-world entities among the software engineers. Formal methods such as VDM and Z come with specification languages which are intended to alleviate the rigour of formalism for using them in software specification. These are often used by the researchers to improve the consistency, traceability and verifiability of the design components created with semiformal methods using UML diagrams.

Sengupta and Bhattacharya [4] proposed a method to ensure consistency between different UML diagrams with the help of a defined set of consistency rules. Z notations and XML are used to analyse different UML diagrams such as class diagram, use case diagram, activity diagram and sequence diagram. The objective of this work is to ensure traceability of requirements in different phases of SDLC. So Z equivalent structure for different UML diagrams is proposed, and consistency between the UML diagrams and their relationship is verified.

Dascalu [3] worked on integration of semiformal, graphical representations with formal notations for construction of time-constrained system. The graphical notations employed are a subset of UML, whereas Z++ is the choice for formal notation. The translation between UML and Z++ is performed in a pragmatic and systematic way with detail algorithm being proposed. It results in a lightweight practical specification which is reliable as well as supports rapid development.

Lausdahl and Lintrup [5] worked on identifying mapping potential between VDM++ and UML diagrams such as class and sequence diagrams. The abstract syntax tree (AST) for VDM is used as an essential part of the model transformation. A tool is built to support bidirectional transformation rules for each construct language. The UML diagrams are exchanged between tools using XML-based standard. This work also constructed a model transformation between sequence diagram and VDM++ traces.

Mota et al. [6] observed that graphical specifications such as UML need to be formally verified, before the implementation phase, in order to guarantee the development of more reliable systems. This work presents a protocol interface for joining computer-aided software engineering (CASE) using UML and formal verification techniques (FVT). It uses automatic property extraction from UML diagrams and first-order logic (FOL)-based level of suitable mechanisms for keeping track of the aspects of system development which are verified.

Most of these works discussed above depend heavily on the model of transformation between formal and semiformal techniques but lack in any commonly acceptable algorithm. In this paper, we focus on independent construction of semiformal and formal specifications, and instead of building a complete transformation model, first we define the consistency rules to check the continuity of different UML constructs with respect to VDM-SL specification, and secondly, we propose a verification model that takes both semiformal and formal specifications as input and results in tracing back the design article with the requirement specified in the requirement analysis phase. We use ILMS as a case study to illustrate our proposal specially because the requirement analysis and design of e-Learning software is a challenging job considering the diversity of its users, standards and models followed in education [7].

3 Proposed Framework

Figure 1 represents the framework for the proposed methodology. The objective of this framework is to provide a systemic approach for guiding the requirements written in NL into a correct and implementable software design specification using requirement model and design model. A systematic approach in requirement engineering (RE) helps in discovering and understanding the requirements at different levels of abstraction and also makes them traceable and verifiable early in

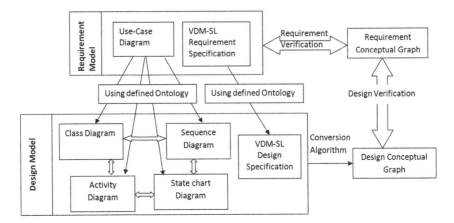

Fig. 1 Proposed framework

the project. It also brings confidence to the design process about the correct implementation of the requirement. The requirement model (Fig. 1) represents the individual requirement with the help of use case diagram and corresponding VDM-SL specification. Requirements being initially available or specified in NL statements are thus hard to verify against the requirement model. In general, conceptual graph defined by Sowa [8] can be used to model NL sentences in a formal and yet easily understandable way. So we convert the requirement in NL statements into conceptual graph and use it for the verification of the requirement model. We will not discuss the requirement modelling in this paper in detail, but we take the output of the requirement model that comes in the form of use case and VDM-SL specification which is already verified against the requirements available in NL statements. Our focus in this work is rather on the design model and its verification.

We consider individual requirement representation in requirement model by use case diagrams that are elaborated in the design phase by means of class diagram, sequence diagram, activity diagram and state chart diagram, each of them representing different perspectives of modelling. The transition from requirement model to design model involves elaboration of the requirement concepts ("what to be done") towards solution of the problem ("how to be done"). Such transition involves many assumptions on the domain, so an early identification and decomposition of the structure of the objects to be used are required to be specified. Our defined ontology serves this purpose; it represents the basic hierarchical structure of the components or terms in the form of object, process and entity. This ontology works as a common agreement between the semantics of the components and is used throughout the RE and design phases. The VDM-SL design specification is the extension of the VDM-SL requirement specification where the operations and data types are elaborated in accordance with the ontology. Figure 2 shows the partial ontology for ILMS case study.

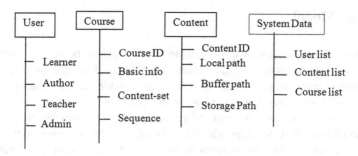

Fig. 2 Partial ontology for ILMS

4 The Case Study

We illustrate the proposed methodology with a case study from the domain of intelligent learning management system (ILMS). ILMS fits in conjunction with intelligent tutoring system (ITS) and conventional LMS. It grasps the essence of ITS in terms of adaptivity within the context of LMS. We consider the following sample requirements as case study. We choose only some of basic functionalities to keep the illustration simple.

> Authors upload contents. The LMS agent manages it in repository. Teachers create courses from the available contents. The system agent manages the content and the course in the repository.

5 Requirement Specification

The requirement analysis phase will redefine the NL requirements, and it represents it using use case diagram, VDM-SL specification and conceptual graph. The output of the requirement model is available for a simple requirement as follows:

R1: *Author creates content* is represented as shown below:

Figure 3.

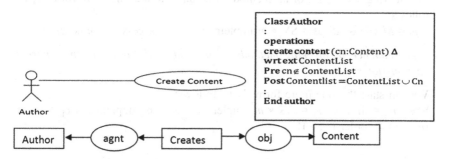

Fig. 3 Requirement specification

6 Design Model

Although many researches have been carried out [9–11] in the field of automatic derivation of class, sequence, state chart and activity diagrams from use case requirements, unfortunately most of them can provide partial benefits to the requirement analysis and design models. In this paper, we use manual methods of translating use cases into class, activity and state chart diagrams which are easily understandable by both developers and domain experts. The design model (Fig. 1) consists of different UML components such as class diagram, activity diagram and state chart diagram. These UML diagrams are derived from the use case with the help of the ontology and VDM-SL specification of the requirement model. However, a continuity checking between the diagrams is essential to ensure consistency between the two models (requirement analysis and design) of software development life cycle. As the UML diagrams and VDM-SL specification used to design the functionalities of the requirements are disjoint in nature, hence, we propose a set of consistency rules that must be satisfied by the design components to ensure continuity between different UML diagrams and VDM-SL.

6.1 Consistency Rules

i. *Each action state of activity diagram corresponds to a use case in the use case diagram.*

   Considering AS as a set of all action states and UC as a set of all use cases, we can state the rule more formally as follows: $\forall e \in AS \; \{\exists u \in UC \bullet e \text{ corresponds } u\}$

ii. *Each action state must access state variables mentioned as constrained.*

   Considering SV as a set of all state variables, we can state the rule more formally as follows:
   $\forall e \in As \; \{\exists s \in SV \bullet e \text{ access } s\}$

iii. *Methods implementing an action state must access its variables.*

   Considering M as a set of all methods, we can state the rule more formally as follows:
   $\forall m \in M \; \{\exists e \in AS \; \exists s \in SV \bullet m \text{ implements } e \wedge e \text{ access } s \wedge m \text{ access } s\}$

iv. *The sequence of invoking methods should match the order of their parent action states.*

   We can state the rule more formally as follows:
   $\forall m_i, m_j \in M \; \{\exists e_i, e_j \in AS \bullet m_i \text{ implements } e_i \wedge m_j \text{ implements } e_j \wedge \text{ order } (m_i, m_j) \cong \text{ order } (e_i, e_j)\}$

v. *The variable used by design-level VDM should have a correspondence with the requirement-level VDM* via *the ontology structure.*

Considering DVDM as a set of all state variables used in design specification of VDM and RVDM as a set of all state variables used in requirement specification of VDM and ONT as a set of all nodes in the ontology, we can state the rule more formally as follows:

$\forall v_i \in$ DVDM, $v_j \in$ RVDM $\{\forall v_a, v_b \in$ ONT • m_i implements $e_i \wedge m_j$ implements $e_j \wedge$ hierarchy $(v_i, v_j) \cong$ hierarchy $(v_a, v_b)\}$

6.2 Use Case Diagram

Figure 4.

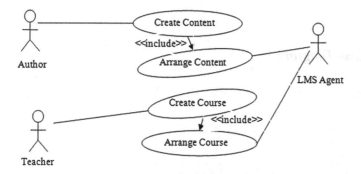

Fig. 4 Use case diagram of case study

6.3 Activity Diagram

Figure 5.

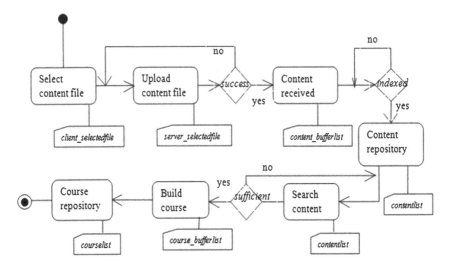

Fig. 5 Activity diagram of case study

6.4 Class Diagram

Figure 6.

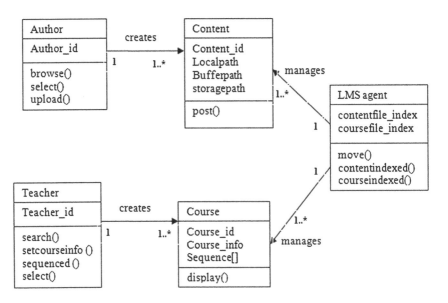

Fig. 6 Class diagram of case study

6.5 Sequence Diagram

Figure 7.

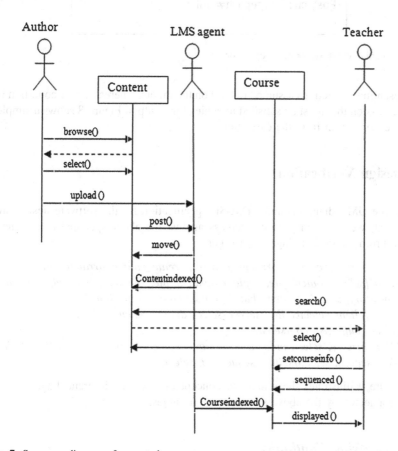

Fig. 7 Sequence diagram of case study

6.6 VDM-SL Specification

The vienna development method specification language (VDM-SL) is a well-established formalizing tool for requirements and design specification [12]. VDM-SL follows a mathematical model based on simple algebraic theory and logic and specifies system's behaviour in its required level of abstraction. VDM-SL expresses system behaviour as logic expressions in terms of operations [13]. The proposed approach focuses only on operations which are defined with the help of pre- and post-conditions. A pre-condition is an expression over the input variables

```
:
operations
move (cn:Contentlist) Δ
Pre cn.storagepath= nil ∧ cn.bufferpath≠ nil
Post cn.storagepath≠ nil
:
```

Fig. 8 Partial VDM-SL for move operation

representing restrictions assumed to hold on the inputs, whereas a post-condition is an expression that must be satisfied to achieve the output. Figure 8 shows a simple "move" operation from the case study.

7 Design Verification

After the UML diagrams and VDM-SL specification for the software design are prepared, we will verify its correctness against the above-specified consistency rules. From the design diagrams, we get

$UC = \{create\_content, arrange\_content, create\_course, arrange\_course\}$

$AS = \{select\_content\_file, upload\_content\_file, content\_received, content\_repository, search\_content, build\_course, course\_repository\}$

$SV = \{client\_selectedfile, server\_selectedfile, content\_bufferlist, contentlist, cousre\_bufferlist, courselist\}$

$M = \{browse, select, upload, post, move, contentindexed, courseindexed, display, search, setcourseinfo, sequenced, select\}$

Figure 9 illustrates the schematic concept of how the different diagrams are interconnected by the above-specified set variables.

7.1 Verifying Continuity

Let us now check the correctness of the design artefacts against the specified consistency rules. For the sake of simplicity, we will discuss only the first three rules with respect to the case study. Table 1 shows mapping between diagrams for the case study.

Rule I: Every action state in $(e1..e7)$ has a corresponding origin in use cases $(uc1..uc4)$.

Rule II: Each action states accessed at least one state variable (from Fig. 5).

Rule III: Method $m5$ accessed the variable *contentlist* which belongs to e4, while m5 implements e4 (from Fig. 8).

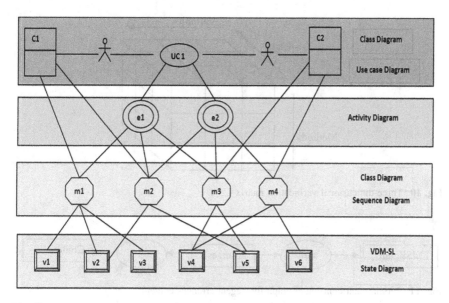

Fig. 9 Interconnection between design components

Table 1 Continuity mapping between UML diagrams

Use case	Action states	Methods	Class
Create content [UC1]	Select content [e1]	Browse [m1]	Author
		Select [m2]	
	Upload content [e2]	Upload [m3]	
Arrange content [UC2]	Content received [e3]	Post [m4]	Content
	Content repository [e4]	Move [m5]	LMS agent
		Indexed [m6]	
Create course [UC3]	Search content [e5]	Search [m7]	Teacher
	Build course [e6]	Courseinfo [m8]	
		Selectcontent [m9]	
		Sequenced [m10]	
Arrange course [UC4]	Course repository [e7]	Indexed [m11]	LMS agent
		Displayed [m12]	Course

7.2 Verifying Requirement

Verifying consistency rule gives us only partial view about the correctness of the design; for a complete view, the design artefact should be checked with the requirement specification. Let us now check whether invoking $m5$ is justified against UC2. In other words, we will check that "agent moves content" is the correct design article of the requirement "agent arranges content".

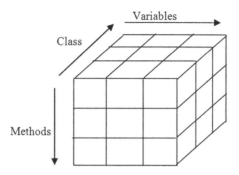

Fig. 10 Three-dimensional verification matrix

Fig. 11 Requirement conceptual graph for "agent arranges content"

The proposed verification technique (Fig. 10) is based on a three-dimensional verification matrix where the three dimensions are methods, classes and state variables. The data in each cell are either 0 or 1, representing the absence or presence of the data variable belonging to the corresponding dimension.

In the case study, *move* () uses two variables *storagepath* and *bufferpath*, so value corresponding to agent class and content class will be one. At the same time, we can map the method $m5$ with the use case UC2 from Table 1. So we can write this formally as "*invoking m5*" ≅ "*Agent, arrangecontent, content*". Now, manually we can check that "*Agent, arrangecontent, content*" is equivalent to the requirement conceptual graph in Fig. 11.

8 Conclusions

This work proposes a methodology to bridge the semantic gap between requirement specification and design artefacts. The success of the design model depends on the correctness of the requirement model. We assumed that the requirement specification used combination of use case, VDM-SL and conceptual graph. The design model illustrates the use case into activity, sequence and class diagrams. As the semiformal UML diagrams and the formal VDM-SL specification are disjoint in nature, we proposed a verification technique to check the continuity between the design artefacts. We also proposed a three-dimensional-matrix-based verification of the design component with the requirement specification. The use of formal method such as VDM and the availability of UML-based CASE tools enhance the possibilities of making the proposed verification method automated. This can be the future extension of this work.

References

1. Booch, G., Rumbaugh, J, Jacobson, I.: The Unified Modeling Language User Guide. Pearson Education India, New Delhi (1999)
2. OMG: Unified modeling language specification, version 2.0. Available at http://www.omg.org/uml
3. Dascalu, S.M.: combining semi-formal and formal notations in software specification: an approach to modelling time-constrained systems. PhD thesis. Department of Computer Science, Dalhousie University, Halifax, Sept 2001
4. Sengupta, S., Bhattacharya,S.: Formalization of functional requirements and their traceability in uml diagrams—A Z notation based approach. In: Proceedings of the 11th Systems Engineering Test and Evaluation Conference (SETE'06), Melbourne, Australia, 25–27 Sept 2006
5. Lausdahl, K.G., Lintrup, H.K: Coupling overture to MDA and UML. Overture Workshop, Newcastle (2009)
6. Mota, E., Clarke, E., Groce, A., Oliveira, W., Falcao, M., Kanda, J.: VeriAgent: an approach to integrating UML and formal verification tools. Electron. Notes Theor. Comput. Sci. **95**, 111–129 (2004)
7. Sengupta, S., Dasgupta, R.: Identifying, analysing and testing of software requirements in learning management system. In: Proceedings of 7th International Conference on Virtual Learning (ICVL) (2012)
8. Sowa, J.: Conceptual graphs: draft proposed american national standard, conceptual structures: standards and practices. Lecture Notes in Computer Science, vol. 1640, pp. 1–65 (1999)
9. Yue, T., Briand, L.C., Labiche, Y.: An automated approach to transform use cases into activity diagrams, modelling foundations and applications. Lecture Notes in Computer Science, vol. 6138, pp. 337–353 (2010)
10. Liwu L.: A semi-automatic approach to translating use cases to sequence diagrams. In: Proceedings of Technology of Object-Oriented Languages and Systems, pp. 184–193, Jul 1999. doi:10.1109/TOOLS.1999.779011
11. Liwu, L.: Translating use cases to sequence diagrams. In: Proceeding of ASE '00, 15th IEEE International Conference on Automated Software Engineering, p. 293
12. Sengupta, S., Dasgupta, R.: Integration of functional and interface requirements of an web based software: a VDM based formal approach. In: Proceeding of IASTED International Conference on Software Engineering (2013). doi:10.2316/P.2013.796-017
13. Larsen, P.G., Battle, N., Ferreira, M., Fitzgerald, J., Lausdahl, K., Verhoef, M.: The overture initiative–integrating tools for VDM. ACM Softw. Eng. Notes **35**(1), Jan 2010

A Lightweight Implementation of Obstruction-Free Software Transactional Memory

Ankita Saha, Atrayee Chatterjee, Nabanita Pal, Ammlan Ghosh and Nabendu Chaki

Abstract Software transactional memory (STM) has evolved as an alternative for traditional lock-based process synchronization. It promises greater degree of concurrency and faster execution. This paper proposes a simple, lightweight, and yet efficient implementation of OFTM. The major contribution of the paper is in proposing a new STM algorithm that uses simple data structure. This does not require any contention manager toward ensuring progress condition, atomicity, and serializability of transactions besides maintaining data consistency. Experimental simulation on random data set establishes the merit of the proposed solution.

Keywords Concurrency control · Obstruction freedom · Abort freedom · Throughput · CPU cycle · Spin count

1 Introduction

Transactional memory (TM) is a concurrency control mechanism to exploit parallelism in modern multiprocessor environment. A transaction in TM executes series of reads and writes to shared memory by executing an atomic block of code.

A. Saha · A. Chatterjee · N. Pal · A. Ghosh · N. Chaki (✉)
Department of Computer Science and Engineering, University of Calcutta, Kolkata, India
e-mail: nabendu@ieee.org

A. Saha
e-mail: arnabianki@gmail.com

A. Chatterjee
e-mail: atrayeechatterjeeiii@gmail.com

N. Pal
e-mail: nabanita1209@gmail.com

A. Ghosh
e-mail: ammlan.ghosh@gmail.com

© Springer India 2015
R. Chaki et al. (eds.), *Applied Computation and Security Systems*, Advances in Intelligent Systems and Computing 305, DOI 10.1007/978-81-322-1988-0_5

TM provides an alternative to the traditional lock-based process synchronization, where program can wrap its code in a transaction. Herlihy and Moss were first to propose hardware-supported TM in 1993 [1] to ensure the consistency of data when shared among several processes. In 1995, Shavit and Touitou [2] coined the term (STM) to describe software implementation of TM for multiword synchronization on a static set of data. The STM implementation is a non-blocking synchronization construct where processes do not need to wait for accessing concurrent objects during contention: a concurrent process either aborts its own atomic operation or aborts the conflicting process. The non-blocking synchronization offers three different types of progress guarantees [3]: wait freedom [4], lock freedom [5], and obstruction freedom [6]. The wait freedom guarantees that every transaction will complete in a finite number of steps. In lock freedom, some transactions will complete in a finite number of steps. And the obstruction freedom demands that every transaction will commit in the absence of any contention. On the basis of progress guarantee, the wait freedom is the strongest and the obstruction freedom is the weakest non-blocking implementation. Even if obstruction freedom is the weakest, its simplicity and faster performance have made an increasing interest among the researchers. All non-blocking synchronizations are free from deadlock, priority inversion, and convoying problems. However, obstruction-free transactional memory (OFTM) may face live-lock problem if a group of processes keep preempting or aborting each other's atomic operations. (DSTM) [7] is an OFTM implementation that minimizes the live-lock problem by implementing various back-off techniques of contention management policies. In DSTM, when a transaction faces contention, it aborts the conflicting transaction or back-off for some specific time to give a chance so that the conflicting transaction can commit. The decision, whether to abort or back-off, is been taken by consulting the contention manager. There are several contention management policies to resolve the contention among transactions [8].

There are other OFTM implementations [9–11] that present an improved solution of obstruction-free non-blocking synchronizations. ASTM [9] offers adaptive methodology to adjust the object acquire scheme in read-dominated and write-dominated workload. In lazy acquire scheme, transaction acquires the data at commit time, and in eager acquire scheme, transaction acquires the data earlier and detects contention earlier. Thus, ASTM uses eager acquire scheme in write-dominated workload and lazy acquire scheme in read-dominated workload. ASTM increases the throughput by this adaptive nature of object acquire methodology. Non-blocking zero indirection transactional memory (NZTM) [12] obeys the same obstruction-free philosophy, but the design is considerably different from that of DSTM and ASTM and relies much more on the underlying hardware architecture. Most importantly, unlike DSTM and ASTM, the NZTM uses in-place data to overcome data indirection overhead. In [11], a transaction is allowed to execute without any abort or back-off during contention. The proposed method uses a modified data structure similar to that of DSTM. In this implementation, transactions consult with contention manager before back-off or self-abort. The implementation does not consider the scenario where multiple transactions share the data object for write operation.

The recent research trends have shown an interest on reducing the abort while ensuring the progress guarantee [13, 14]. In [13], multiple versions of data object are being maintained to avoid spurious abort for read-only transactions. The proposed method in [14] shows that by using single version of data object, abort-free execution for read-only transactions is possible.

In this paper, a new algorithm has been proposed toward developing an OFTM based on some preexisting works [11], where multiple transactions share the same data object and execute in an abort-free manner. A pool of transactions is generated by using existing randomization algorithms such as linear congruence [15]. In the simulation, a more number of write transactions have been considered as compared to the number of read-only transactions. This is strikingly different as TM is known to perform better for read-dominated transactions, and most of the prior works utilize the same while measuring performances. In order to make the proposed STM lightweight, the data structure of TM object has been modified from what has been used in earlier works [11]. The approach does not require contention manager to resolve the conflict among transactions as the execution pattern of transactions itself is capable to resolve the contention if any. Moreover, the proposed method uses a simple validation mechanism to check for consistency.

In Sect. 2, the state-of-the-art progress condition in STM is described. In Sect. 3, the basic concept has been stated. Section 4 presents the proposed lightweight OFTM. In Sect. 5, the performance of the proposed algorithm has been evaluated.

2 Software Transactional Memory and Progress Condition

Transactions, in STM, are dynamic sequence of operations that executes in parallel as a single atomic operation till they do not conflict. STM ensures that execution of transaction will be either successful, in which case it commits by making updates permanent, or unsuccessful, in which case transaction aborts by discarding all its updates. When two transactions run concurrently on same data, at least one of them modifies it, and conflict occurs. STM resolves this conflict by aborting any one of these transactions. Aborted transaction may reinitiate later and commit eventually. Frequent aborts tend to waste system resource and deteriorate the performance. Thus, the objective of STM is to allow as much as transactions to make progress concurrently and commit eventually. This progress condition is termed as positive concurrency. Positive concurrencies are of two types: progressiveness and permissiveness.

Progressiveness is an execution pattern that allows a transaction to commit. It demands that transaction encountering no conflict must always commit. Among a group of conflicting transactions, the progressiveness demands that at least one of the transactions will commit [18]. This is a stronger version over the basic progressiveness. However, STM generally aborts a conflicting transaction to resolve contention. Thus, strong progressiveness is not the strongest one because it cannot guarantee that all the conflicting transaction will eventually commit.

Permissiveness demands that a transaction is never aborted unless it is required for maintaining correctness [19]. The STM systems that randomize transactions' commit/abort point using some random functions are known as probabilistically permissive. In [20], Guerraoui et al. also indicated that some STM systems check that the data consistency at commit point, i.e., check the value of data object read by the transaction, has not been modified, and if modified, then abort the transaction. These types of transactions are not permissive with respect to the opacity condition.

The available permissive STM techniques to detect and resolve conflict are prone to error and susceptible to false abort. Thus, an important goal of permissiveness is to avoid spurious aborts. Multiversion (MV) permissiveness [13] minimizes the rate of aborts for conflicting transactions by segregating read and write transactions. MV permissiveness ensures that read-only transactions will never abort. This is achieved by maintaining multiple versions of each data. As maintaining multiple versions requires additional storage and complex computational mechanism, the PermiSTM [13] is being evolved that supports MV permissiveness by keeping only single version of each data item. Whether MV permissiveness or PermiSTM, none of them guarantee abort freedom for write transactions in the presence of conflict.

The obstruction-free software transactional memory (OFTM) [6] guarantees progress for a transaction when all other transactions are suspended. As per the generic definition given in [6], an obstruction-free synchronization guarantees progress for any thread that eventually executes in isolation. DSTM [7] is the first OFTM implementation that uses a dynamic transactional memory object (TM Object), which contains a pointer to the locator object. The locator object points to the descriptor of the most recent transaction and holds old and new versions of the data object. A transaction may be in active, committed, or aborted state. The descriptor of the transaction holds this state. When a transaction acquires some object for read/write operation, it makes the status filed as 'Active'. When transaction commits successfully, it changes its status filed as 'Committed'. The status field is set as 'Aborted' when a transaction is aborted by other conflicting transactions.

When a transaction, say T_k wants to read an object X and finds that another transaction, say T_m is updating X, then T_k may eventually abort T_m or back-off for specific time after consulting with contention manager. Otherwise, if T_k finds that no other transaction is updating X, then it reads the current value of X, and at commit point, it checks for the consistent state of X. Transaction T_k commits if it finds consistent state of X.

When transaction T_k wants to update the object X, it tries to acquire an exclusive but revocable ownership of X. To do so, T_k gets the information of X. If T_k finds that no other transaction has owned the object X, it exclusively owns the object. Otherwise, if T_k finds another transaction, say T_m is updating X, then the contention manager will decide whether T_k will back-off for a random duration or abort T_m to acquire the ownership of X. All the available OFTM techniques [9–11, 16] follow this basic high-level principle although they differ in the implementation techniques to achieve better throughput. An OFTM guarantees that a transaction commits in the absence of contention. However, it is unable to provide concurrency since progress is guaranteed only when one transaction is active at a time [17].

In [11], algorithm implements concurrency, where the second transaction is allowed to proceed immediately without affecting the execution of the first transaction in the presence of contention. This method uses the basic data structure of DSTM [7] but differs in the execution pattern. When a transaction, say T_x, wants to access a data object, it checks whether the data object is already accessed by another transaction or not. If data object is not accessed by any other transactions, then T_x owns the data object. Otherwise, if it finds that the other transaction, say T_y, owns the data object, then T_x reads the data object value from the T_y's new data field. In such case, when T_x reaches its commit point, it checks whether T_y is committed and the data read by T_x is consistent or not. On failing to satisfy any of the conditions, T_x re-executes its operation. In this algorithm, transactions take the help of contention manager before back-off or abort. This approach yields a higher throughput as compared to DSTM. It [11] supports abort-free execution for both read-only and write transactions. However, the solution is restricted for only two transactions. Like other OFTM approaches, the technique in [11] cannot support permissiveness with respect to the opacity condition as transactions check for consistent state of data object at their commit point. Moreover, the isolation property too is not satisfied as the data object being updated by some transaction, say T_y, is accessed by another transaction, say T_x, before T_y commits. In this paper, lightweight OFTM approach has been proposed. Concurrency is achieved for multiple transactions in a pairwise execution, where a transaction does not abort after it begins to execute. When a transaction ends working with data object, it checks for its previous transaction's data access status (whether still accessing the data object or not) and/or checks for the data inconsistency. The transaction re-executes its data access region (i.e., spins) with the current value of the data object when necessary to resolve the contention. The method does not require additional contention manager.

The basic methodology of the proposed execution is motivated from [11] although it uses a simplified data structure causing lesser number of memory accesses than [11]. In the next section, the scope of the proposed work has been elaborated.

3 Scope of the Work

The objective of the proposed algorithm is to maximize the throughput by reducing the total execution time for a set of transactions. The proposed STM system keeps the information about transactions' data access region, i.e., when a

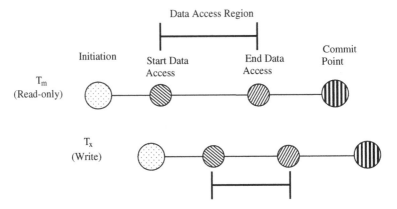

Fig. 1 When T_x reaches its end data access point, T_m is out of its data access region. Thus, T_x commits eventually

transaction starts accessing the object and stops working with that object. The execution flow of a transaction includes initiation point, data access region, and commit point. It is assumed that each transaction has a single data access region. Every transaction is allowed to access data, available in TM object, without detecting any conflict. When a write transaction, say T_x, reaches its end data access point, it checks the transaction type of its previous transaction, say T_m.

- **If T_m is read-only transaction**, then T_x checks whether T_m is out of its data access region (i.e., its data access is completed) or not. If T_m is out of its data access region, then T_x continues its execution and commits eventually. Otherwise, T_x spins for specific time equal to its data access region and checks again whether T_m is out of its data access region or not. In this case, T_x only commits when T_m is out of its data access region. Both the scenarios are depicted in Figs. 1 and 2. In neither case, T_x required to check data inconsistency as its previous transaction T_m is read only and thus not going to update the data object.

- **If T_m is write transaction**, then T_x checks whether T_m is out of its data access region or not. If T_m is out of its data access region (but yet to commit), then T_x reads from T_m's last updated data value and re-executes its operation and commits. Otherwise, when T_x finds that T_m is within its data access region, it spins for specific time equal to its data access region and checks again whether T_m is out of its data access region or not and follows the same procedure as above. After spin T_x may find that T_m is already committed, then T_x accesses the last updated data value of TM object to re-execute its operation and commits. This is illustrated in Figs. 3 and 4.

When T_x and T_m both are read-only transactions, the case is trivial since no contention arises. If T_x is read-only transaction and T_m is write transaction, then T_x spins, till T_m is out of its data access region. When T_m is out of its data access region, T_x reads the last updated value of data object from T_m.

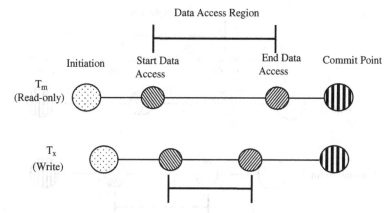

Fig. 2 When T_x reaches its end data access point, T_m is within its data access region. Thus, T_x spins until T_m is out of its data access region

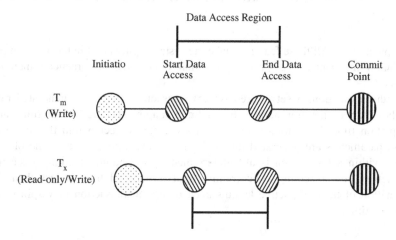

Fig. 3 When T_x reaches its end data access point, T_m is within its data access region. Thus, T_x spins until T_m is out of its data access region

The data structure of the TM object [11] has been modified in the proposed execution for simplicity. Here, the TM object includes the data object only. This reduces the data indirection overhead. Every transaction uses its local memory to store the data value that it reads and updates during its execution. Thus, data storage mechanism has also been simplified. Moreover, the execution does not include the contention manager as the proposed execution pattern is capable to resolve the conflict. Further, each transaction has an update bit which determines whether a transaction is a read-only (update bit = 0) or write (update bit = 1) transaction. Inclusion of this update bit reduces the overhead of data consistency check for a transaction that is preceded by a read-only transaction (as explained in

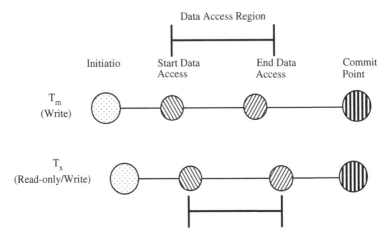

Fig. 4 When T_x reaches its end data access point, T_m is out of its data access region. Thus, T_x reads data value from T_m and re-executes its operation and commits when T_m is successfully committed

example above). All these features make the system lightweight in terms of simple validation mechanism, lower computational overheads, and reduced memory access.

In this execution, a set of (read/write) transactions has been generated randomly. Each of the transactions has a finite length, specific arrival time, and completion time. Each transaction accesses the data object within this interval. These parameters are generated randomly in this execution. Access time of TM object within a transaction is also determined by a random function. Since the transactions are executing concurrently, total turnaround time is less than the summation of their respective lengths even though a transaction may spin more than one time.

4 Simulation of the Proposed Lightweight OFTM

Some important terminologies used in the rest of this paper such as data access region, spin count, and transaction execution length have been defined in this section.

4.1 Definitions

- *Spin count* is the number of times the transaction T_x executes its data access region, while its previous transaction T_m is in its data access region.

Spin count is an overhead of the proposed method. However, it has been observed that the number of transactions undergoing spin is much less with respect to the subset considered in the execution phase.

Therefore, even if the average spin count is high, then also the total execution time never exceeds the time when the subset of transactions are executed serially.

- *Data-access region* (*DR*) is the time interval where (i.e., data access start point to data access end point) each transaction accesses a shared resource.

  Every transaction has a single data access region of a finite length.
- *Length* (*absolute value*) of the transaction is the time limit throughout which the transaction executes.
- *Extra CPU cycle* is the overhead of the CPU caused by unnecessary spinning of a transaction T_x, when both T_x and its preceding read-only transactions are in their respective data access regions.

  Consider that T_m is the preceding read-only transaction for T_x. Transaction T_x keeps on spinning till T_m comes out of its data access region. In terms of data consistency, this is unnecessary as the read-only transaction T_m will never modify the value of the data object. The significance of defining extra CPU cycle lies in this. This is different from computing spin count in case the first transaction T_m is a write transaction.
- *Update bit* (one bit integer) determines whether a transaction is a read-only (0) or write (1) transaction.

4.2 Assumptions

- There will be only one shared data object to be accessed by a set of transactions.
- Proposed method considers both read-only transactions (update bit = 0) and write transactions (update bit = 1).
- Every transaction has a single data access region of finite length.
- It is being assumed that that T_m precedes T_k $\forall m, k \in [1 \ldots n]$ iff transaction T_m occurs before transaction T_k. Moreover, the transaction T_k will exit from its data access region only when transaction T_m is out of its data access region.
- At a particular instance, during execution phase, only two transactions are considered for checking the constraints.
- For better randomness, linear congruence [15, 16] method is used as the random function to generate a pool of transactions.
- The subset of transactions is selected transactions from three regions, i.e., beginning, middle, and end of the transaction pool.

4.3 Algorithm

Procedure lightweight_OFTM
  /* Main procedure that coordinates other sub-procedures to
  fetch the required output
  */
Begin
    Call function lin_cong to generate a pool of
    transactions randomly;
    Store the transaction parameters in *transaction_file*;
    Select a subset of transactions from the above pool;
    Call procedure *STM_execution*;
    Call *Display*;
End

Function lin_cong
  /* Random number generation : The linear congruential method
  [12] produces a sequence of integers X_1, X_2, X_3... between zero
  and *m-1* according to the following recursive relationship:
  X_{i+1} = (a X_i + c) mod m, where i=0,1,2,... and X_0 is called the
  seed, *a* is called the constant multiplier, *c* is the increment
  ,*m* is the modulus.
  */
Begin
    Generate random numbers using expression X_{i+1} = (a X_i +
    c) mod m;
End

Function STM_execution
  /* A subset of transactions run concurrently to access a
  single shared *tm_data* (TM object), satisfying few
  constraints.
  */
Begin
    for *tr=(start+1) to (subset)+start*
            /*Loop runs for each of the transaction in the subset,
            where *start* is the beginning of the subset */
            Set *two pointers* for accessing the previous and
            current transaction;
            Check consistency of the *tm_data;*
            Increment *tm_data* by 1, for each write
            transaction;
            Store the cumulative execution time in
            execution_array;
            Increment no_of_*spin_count* for every current
            transaction, whose previous transaction is
            within the data access region ;
            Increment *extra_cpu_cycle* for every (read/write)
            transaction, whose previous transaction is a
            read transaction (and is within the data access
            region);
        End for
    End

```
Procedure Display
    /* Display all the evaluated values.*/
Begin

    total_ execution_ time= max(execution_array) ;
    /* function max returns largest element from array */

    average_spin_count=no_of_spin_count/tr_spin;
    /* tr_spin= number of transaction undergoing spin*/

    average_total_execution_time=total_execution_time/car
    dinality_of_subset;

    Print new modified value of tm_data,
        no_of_spin_count,
        average_spin_count,
        total_ execution_time,
        average_total_execution_time,
        extra_cpu_cycle;

    Store all the evaluated values in Result_file;

    Call procedure comparison;

End


Procedure comparison
    /* For comparing the theoretical and observed value (of
    execution units).
    */
Begin
    /* Summation of all the length of transactions in the subset
    to obtain the theoretical value.
    */
    Calculate the theoretical execution time;
    Compare theoretical and observed value;

    /* Percentage of reduced execution units*/
    Compute reduce_factor;
End.
```

5 Performance Evaluation by Experiments

The proposed algorithm has been implemented using C language. Based on multiple parameters, a characterization of transaction is done and multiple set of transactions are generated randomly. Performance is measured with respect to multiple set of transactions generated randomly. Experimental verification is stated in Sect. 5.1 for the sake of completeness. This is followed by experimental results (Sect. 5.2).

5.1 Plan for Experimental Verification

The problem domain is divided into three phases:

Phase I (Transaction Generation): Initially, a set of N transactions are generated. It includes the following steps:

(a) 'Number of transactions' has been taken as a user input.
(b) Initially, the value of data object is a set to a numeric constant.
(c) For each transaction, the following calculations are being done:

     i. Length, arrival time, and completion time of each transaction using a predetermined random function.

     ii. Since it is being assumed that there is *only one* shared data object, it is required to determine the time interval for accessing the data object for each transaction. The interval will be within the length of corresponding transaction. Using another random function, different from the function that is used to determine the transaction length, the *lower and upper limits* of accessing the data object are being evaluated. It is also being assumed that each transaction can access the data object *only once*, and thus, range of data access region is calculated only once for each transaction.

(d) Aforesaid generated transactions will be stored in a file. There will be six parameters associated with each transaction

     i. Transaction id (TI)
     ii. Length of transaction (TL)
     iii. Arrival Time of transaction (AT)
     iv. Completion Time of transaction (CT)
     v. Region of Data Access—from Lower Limit (LL) to Upper Limit (UL)
     vi. Update bit (UB)—either 0 or 1

Phase II (Subset of Transaction Formation): In this phase, the subset of transactions has been taken into consideration for further analysis. The number of transactions to form the subset is being taken as user input. Say, the subset contains 'B' number of transactions. This subset can be chosen from any region of the transaction pool.

Phase III (Execution Phase): Let T be the turnaround time for executing B number of transactions, which run concurrently as per the proposed method. Now, we need to show that $T \leq \sum_{i=1}^{B} L_i$ where L_i is the length of the transaction T_i.

In this phase, total execution time, average execution time, total number of spins, average number of spins, and extra CPU cycle are evaluated. These are used to compare between theoretical results when corresponding transactions run sequentially. The sample set for the algorithm is shown in Figs. 5, 6 and 7. Each column represents different parameters (transaction_id, transaction length, arrival time, completion time, lower limit and upper limit of data access region, and update bit) of transactions. Figs. 6 and 7 are two subsets of transactions selected from pool of transactions (Fig. 5) from beginning and end, respectively (with subset cardinality as 3 and 4).

Fig. 5 Set of 10 transactions after phase I

TI	TL	AT	CT	LL	UL	UB
T1	63	46	108	69	95	1
T2	45	49	93	79	90	1
T3	141	61	201	90	192	0
T4	129	152	280	273	279	1
T5	93	202	294	291	292	1
T6	135	244	378	375	377	0
T7	111	339	449	368	439	1
T8	39	414	452	431	444	1
T9	123	416	538	532	537	0
T10	75	460	534	514	531	1

5.2 Experimental Results

The lightweight property of the proposed algorithm is derived from the fact that it has reduced the number of memory accesses in comparison with [11]. This is achieved by simplifying the data structure. TM object includes data object, and thus, the data indirection overhead is reduced. Secondly, each transaction can access the data object (available at TM object) and stores the read value of the data object in its local variable. The proposed method does not have the descriptor's field such as 'Status', OldData, and NewData as it was in [11]. Besides, in [11], every transaction checks the data consistency in its commit point. In case of inconsistent data, transaction re-executes its operation from the data access point to the commit point [11]. In the proposed method, the system has the data access information a priori. When a transaction reaches the end of data access region, it checks whether its previous transaction is out of its data access region or not. If the previous transaction is a read-only one and is out of its data access region, then the current transaction commits directly. However, when preceded by a write transaction, the current transaction re-executes its operation after reading the last updated value by previous transaction and then commits.

Contention arises when more than one transactions access the same data object concurrently and at least one of them is a write transaction. In the absence of contention, a read-only transaction accesses the memory only once, i.e., at the time of reading the data value. On the other hand, a write transaction accesses the memory twice to read and update data. In the presence of contention, the number of memory access for the proposed algorithm is computed for the best-case and worst-case scenarios.

In the best-case scenario, a write transaction T_x reaches its end data access point and finds its predecessor, say T_m, is a read-only transaction and is out of its data access region. In this case, both T_m and T_x will commit eventually without facing any contention. Thus, the number of memory access is **3** (1 for read-only transaction and 2 for write transaction). Now, in the same scenario, for **n** number of transactions, where a write transaction is preceded by all read-only transactions the number of memory accesses $= \{(n-1) + 2)\}$.

In the worst-case scenario, a write transaction T_x reaches its end data access point and finds its predecessor, say T_m, is a write transaction and is within its data access region. Thus, T_x spins, until T_m is out of its data access region. When T_x spins, it does not access the data object; thus, the number of memory count remains unchanged. When T_x finds that T_m is out of its data access region (or committed in between T_x's spin), T_x accesses the data object and updates the value at its commit point. Thus, the number of memory accesses is **5** (2 for T_m and 3 for T_x). In the same scenario, for **n** number of write transactions that are executing concurrently, the number of memory access = $\{2 + (n-1)*3\}$.

Although the proposed algorithm has been implemented pairwise, the experimental logic can be easily extended for a set of **n** transactions, both read only and write. In pairwise execution, when a transaction is preceded by a read-only transaction, data inconsistency cannot occur. Hence, no additional memory access is actually required at the end data access region to check for data consistency. The scenario is not the same for a set of **n** transactions because a transaction may be preceded by write transactions earlier even if it is immediately preceded by a read-only transaction. Hence, every transaction, irrespective of whether it is preceded immediately by a write transaction or by a read-only transaction, requires data consistency check at the end data access region. Thus, the number of memory accesses for a read-only transaction is 2 at the start and end of its data access region and 3 for a write transaction with an additional access for updating data object.

In the event where **R** number of read-only transactions and **W** number of write transactions, in a set of n transactions, executing concurrently, the total number of memory accesses is $\{1 + (R-1)*2 + W*3\}$ [when the first transaction is read only] or $\{2 + (W-1)*3 + R*2\}$ [when first transaction is write]. In [11], when a transaction, say T_x, initiates and finds that another write transaction, say T_m, is already owned the data object, then T_x reads from T_m's new data and start executing. At commit point, T_x checks whether T_m has been committed or not. If T_m is active, then T_x again reads data value from T_m and re-executes it operation. Thus, for every spin, T_x requires the memory access. Hence, for a set of **n** transactions, where **R** is the number of read-only transactions and W is the number of write transactions [i.e., $R + W = n$], the number of memory accesses in [11] is $\{R + W*2 + total\_no\_of\_spin\}$. Whenever this spin count increases, the number of memory access is increased.

Thus, the total number of memory accesses for the proposed algorithm is smaller as compared to [11].The efficiency and performance improvement of the proposed method show that it is always preferable to run the transactions concurrently rather than sequentially. The performance improvement criterion is further supported by the results presented in the charts below. The bottom line for the proposed approach would guarantee at least 30 % time save as compared to the theoretical approaches that run transactions serially.

Figure 8 is a comparison of total execution time taken by the transactions when executed serially and the value obtained using the proposed method (i.e., when run concurrently). X-axis denotes the number of transactions that we select for analysis, and y-axis denotes the execution time unit(s). The notation $p(q)$ represents that 'p' number of transactions are selected from a set of 'q' transactions.

TI	TL	AT	CT	LL	UL	UB
T1	63	46	108	69	95	1
T2	45	49	93	79	90	1
T3	141	61	201	90	192	0

Fig. 6 Data set 1 (after phase II)

TI	TL	AT	CT	LL	UL	UB
T7	111	339	449	368	439	1
T8	39	414	452	431	444	1
T9	123	416	538	532	537	0
T10	75	460	534	514	531	1

Fig. 7 Data set 2 (after phase III)

Fig. 8 Improvement in execution time using proposed algorithm

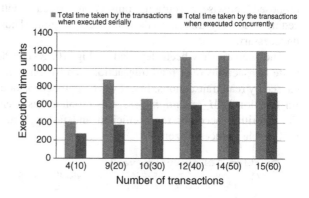

Fig. 9 Spin count versus number of transactions undergoing spin

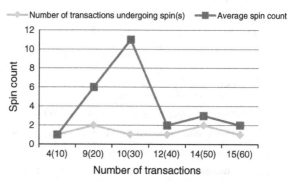

In Fig. 9, x-axis denotes the number of transactions that we select for analysis and y-axis denotes the number of transactions under spin count. Similar to Fig. 8, the notation $p(q)$ represents that 'p' number of transactions are selected from a set

Fig. 10 CPU overhead due
to read transactions

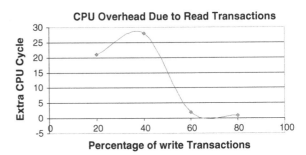

of 'q' transactions. Spin count is obviously an overhead. However, it has been observed that the number of transactions undergoing Spin is much less with respect to the subset considered in the execution phase. Therefore, even for very high value of the average spin count, total execution time never exceeds the time when the subset of transactions are executed serially.

A higher number of read transactions incur greater CPU overhead. In the proposed pairwise execution, transactions execute without checking their preceding read-only transaction's data access region. Thus, transactions re-execute unnecessarily.

The scenario has been depicted in Fig. 10. *X*-axis denotes the percentage of write transactions over read transactions, and *y*-axis denotes the extra CPU cycle caused by read transactions. In a revised approach, as documented in the algorithm for function *STM_execution_ver2*, this problem has been resolved. Consequently, the algorithm becomes more efficient and faster. With this improvement, extra CPU cycle reduces to zero.

```
Function STM_execution_ver2
    /* A subset of transactions run concurrently to access a
    single shared tm_data (TM object), satisfying few
    constraints.
    */
Begin
    for tr=(start+1) to (subset)+start
        /* Loop runs for each of the transaction in the subset,
        where start is the beginning of the subset
        */

        Set two pointers for accessing the previous and
        current transaction;

        Check consistency of the tm_data;

        Increase tm_data by 1 for each write transaction;

        Store cumulative execution time in execution_array;

        Increment no_of_spin_count for every current
        transaction, whose previous transaction is in data
        access region and is not a read transaction;

    End for
End
```

6 Concluding Remarks

An efficient implementation of STM without involving contention manager and by using a simple data structure makes the proposed algorithm suitable for adopting as compared to some of the existing ones. In the proposed method, a transaction, read only or write, is never aborted after its initiation. This ensures progress guarantee for every transaction in the set of n finite-length transactions, n being a finite integer. Lightweight property of the algorithm is justified by avoiding the indirection overhead, reducing the number of memory accesses, simplifying the validation mechanism, and thus reducing the computation overheads. By avoiding all these complex issues, the proposed algorithm has achieved all the basic deliverables of OFTM.

Although the proposed work has been implemented by comparing the transactions pairwise, this comparison can be easily extended for a set of n transactions, where every transaction will check for the data consistency in its data access region if the immediate previous transaction is out of its data access region. The work may further be extended toward an empirical evaluation of the proposed STM for a multicore environment with varying number of cores. Also, there could be more than one TM object for contention within a given set of transactions. In this paper, linear congruence has been used to generate random data sets for different transactions, their length, time and order of occurrences, etc. This may also be achieved using randomization functions, such as Mersenne Twister, WELL ('Well Equi-distributed Long-period Linear'), Blum Blum Shub, and Fortuna.

References

1. Herlihy, M., Moss, J.E.B.: Transactional memory: architectural support for lockfree data structures. In: Proceedings of 20th Annual International Symposium on Computer Architecture, ISCA '93, pp. 289–300, May 1993
2. Shavit, N., Touitou, D.: Software transactional memory. In: ACM SIGACT-SIGOPS Symposium on Principles of Distributed Computing, pp. 204–213. ACM August 1995
3. Marathe, V.J., Scott, M.L.: A Qualitative Survey of Modern Software Transactional Memory Systems. Technical Report Nr. TR 839. University of Rochester, Computer Science Department (2004)
4. Herlihy, M.: Wait-free synchronization. TOPLAS: ACM Trans. Program. Lang. Syst. 13(1), 124–149 (1997)
5. Fraser, K.: Practical lock freedom. PhD Dissertation, Cambridge University Computer Laboratory (2003)
6. Herlihy, M., Luchangco, V., Moir, M.: Obstruction-free synchronization: double-ended queues as an example. In: Proceedings of the 23rd International Conference on Distributed Computing Systems, pp. 522–529 (2003)
7. Herlihy, M., Luchangco, V., Moir, M., Scherer III, W.N.: Software transactional memory for dynamic-sized data structures. In: 22nd Annual ACM Symposium on Principles of Distributed Computing, pp. 92–101, July 2003

8. Scherer III, W.N., Scott, M.L.: Advanced contention management for dynamic software transactional memory. In: 24th Annual ACM Symposium on Principles of Distributed Computing, PODC '05, pp. 240–248 (2005)

9. Maranthe, V.J., Scherer III, W.N., Scott, M.L.: Adaptive software transactional memory. In: Proceedings of the 19th International Symposium on Distributed Computing (DISC), pp. 354–368, May 2005

10. Tabba, F., Wang, C., Goodman, J.R., Moir, M.: NZTM: non-blocking zero-indirection transactional memory. In: Proceedings of the 21st ACM Annual Symposium on Parallelism in Algorithms and Architectures (SPAA), pp. 204–213 (2009)

11. Ghosh, A., Chaki, N.: Design of a new OFTM algorithm towards abort-free execution. In: 9th International Conference, ICDCIT 2013, pp. 255–266, Bhubaneswar, India, 5–8 Feb 2013

12. Harris, T., Larus, J., Rajwar, R.: Transactional Memory, 2nd edn., pp. 101–145. Morgan & Claypool, (2010)

13. Perelman, D., Fan, R., Keidar, I.: On maintaining multiple versions in STM. In: Proceedings of the 29th ACM SIGACT-SIGOPS Symposium on Principles of Distributed Computing, PODC '10, pp. 16–25 (2010)

14. Attiya, H., Hillel, E.: Single-version STMs can be multi-version permissive. In: Proceedings of the 12th International Conference on Distributed Computing and Networking, ICDCN'11, pp. 83–94, Bangalore, India (2011)

15. http://www.eg.bucknell.edu/~xmeng/Course/CS6337/Note/master/node40.html (2014)

16. Knuth, D.E.: The art of computer programming. Seminumerical Algorithms, vol. 2, 3rd edn. Addison-Wesley, Reading (1997). ISBN 0-201-89684-2

17. Guerraoui, R., Kapałka, M.: On obstruction-free transactions. In: Proceedings of the 29th Annual Symposium on Parallelism in Algorithms and Architectures, pp. 304–313 (2008)

18. Guerraoui, R., Kapalka, M.: The semantics of progress in lock-based transactional memory. In: POPL '09, pp. 404–415 (2009)

19. Guerraoui, R., Henzinger, T.A., Singh, V.: Permissiveness in transactional memories: In: Proceedings of the 22nd International Symposium on Distributed Computing (2008)

20. Crain, T., Imbs, D., Raynal, M.: Read invisibility, virtual world consistency and probabilistic permissiveness are compatible. In: Algorithms and Architectures for Parallel Processing, pp. 244–257. Springer, Berlin (2011)

Part II
Cryptography

Multiplicative Polynomial Inverse Over GF(7³): Crisis of EEA and Its Solution

J.K.M. Sadique Uz Zaman and Ranjan Ghosh

Abstract The multiplicative polynomial inverses of all elemental polynomials exist under each of all irreducible polynomials over the finite field $GF(p^m)$ where p is a prime integer and both p and $m \geq 2$. For $GF(2^8)$, the Extended Euclidean Algorithm (EEA) successfully finds multiplicative inverses of all the 255 elemental polynomials under each of 30 irreducible polynomials. However, for $GF(7^3)$, the same algorithm cannot find multiplicative inverses of all the 342 elemental polynomials under each of its 112 monic irreducible polynomials. A simple algebraic method proposed in the paper finds all the 112 monic irreducible polynomials over $GF(7^3)$ along with the multiplicative inverses of all the 342 elemental polynomials under each of the 112 irreducible polynomials.

Keywords Extended euclidean algorithm · Extension field · Galois field · $GF(7^3)$ · Monic irreducible polynomial · Multiplicative inverse

1 Introduction

A finite field, also known as a Galois field $GF(q)$, contains a finite number of elements from 0 to $q - 1$, where q is an integer and is called the order of the field. The finite field can also be denoted by $GF(p^m)$ where p is a prime number and m is a positive integer. When $m = 1$, the field is called a prime field and for $m \geq 2$, the field is called an extension field which is related to polynomials. When $p = 2$, the field is called a binary field.

J.K.M. Sadique Uz Zaman (✉) · R. Ghosh
Department of Radio Physics and Electronics, University of Calcutta, 92 A.P.C. Road, Kolkata 700 009, India
e-mail: jkmsadique@gmail.com

R. Ghosh
e-mail: rghosh47@yahoo.co.in

© Springer India 2015
R. Chaki et al. (eds.), *Applied Computation and Security Systems*, Advances in Intelligent Systems and Computing 305, DOI 10.1007/978-81-322-1988-0_6

Using Extended Euclidean Algorithm, one can calculate the multiplicative inverse of an element $a \in \text{GF}(q)$ over a finite field, except 0, if and only if a is relatively prime to q [1, 2]. For a prime field, the same algorithm calculates the multiplicative inverses for all elements $a \in \text{GF}(p)$. For an extension field ($q = p^m$), there exists elemental polynomials of all elements $a \in \text{GF}(p^m)$ with highest degree $(m - 1)$ or less and a number of irreducible polynomials with highest degree m. The multiplicative inverse of an element is calculated through its multiplicative polynomial inverse obtained from its elemental polynomial under an irreducible polynomial. Considering binary field ($q = 2^m$), one can use the Extended Euclidean Algorithm (EEA) to calculate the multiplicative inverses of all its elemental polynomials under an irreducible polynomial [3, 4]. Following the said algorithm, the multiplicative inverses of 255 elements over $\text{GF}(2^8)$ can be calculated using the first irreducible polynomial among 30 such polynomials and this array of multiplicative inverses is used as a preliminary substitution box to form the S-Box used in AES [5, 6].

The $\text{GF}(p^m)$ has p^m elements, and the polynomial representation $E(x)$ of each the element involves m terms with maximum degree $(m - 1)$. The coefficients a_i belong to $\text{GF}(p)$ and vary from 0 to $(p - 1)$. The expression of $E(x)$ is written as,

$$E(x) = \left\{ \sum_{i=0}^{m-1} a_i x^i : a_i \in \text{GF}(p) \right\}$$

The irreducible polynomial $I(x)$ under $\text{GF}(p^m)$ is also expressed as [7–9],

$$I(x) = \left\{ \sum_{i=0}^{m} a_i x^i : a_i \in \text{GF}(p) \quad \text{and} \quad a_m \neq 0 \right\}$$

The leading coefficient a_m in irreducible polynomials cannot be zero and other coefficients a_i vary from 0 to $(p - 1)$. For $\text{GF}(2^m)$, all irreducible polynomials are monic, since the unity value is the only option for the leading coefficient, while for $\text{GF}(p^m)$ with $p > 2$, the leading coefficient has options between 1 and $(p - 1)$. For such Galois field, there exists monic as well as non-monic irreducible polynomials. When adding two elemental polynomials, the coefficients of elements having identical power are added with modulo p. When two such elemental polynomials are multiplied, the coefficients of elements having identical power are added with modulo p following the addition rule without looking into the fact that the power of x is becoming more than $(m - 1)$. $I(x)$ then modulo operate the product and the result is in the range of polynomials earmarked for $E(x)$.

For $\text{GF}(7^3)$, no information is available with the authors in respect of finding multiplicative inverse using EEA. We adopted EEA to find multiplicative inverse under one of the 112 monic irreducible polynomials over $\text{GF}(7^3)$ with limited success. Its successful and unsuccessful applications are presented in Sect. 2. In Sect. 3, we have proposed a simple algebraic method to calculate multiplicative inverse over $\text{GF}(7^3)$ with the same examples shown in Sect. 2. The computational algorithm of the proposed method is presented in Sect. 4. In Sect. 5, we have presented the results, discussion, and future scopes of the work. The conclusion is in Sect. 6.

2 EEA and Multiplicative Inverse in GF(2^8) and GF(7^3)

There are two algorithms in respect of finding the GCD of two positive integers, A and B for $A \geq B$. Euclidean Algorithm is the one, which finds the GCD directly, while the EEA finds the same through two other integers S and T. Euclidean Algorithm was invented around 300 B.C. by Euclid, a Greek mathematician. During first half of seventeenth century, Bachet, a French mathematician first conceived the idea of Extended Euclidean Algorithm.

The EEA is narrated in Sect. 2.1 along with examples and its computational algorithm. In Sect. 2.2, the EEA is successfully applied to find multiplicative inverse (MI) of elemental polynomial (EP) over GF(2^8) using one of the 30 irreducible polynomials (IP). However, it is noticed that for GF(7^3), the EEA is selective for few elemental polynomials barring few others. A case of successful application of EEA over GF(7^3) is shown in Sect. 2.3, while Sect. 2.4 depicts an unsuccessful application. A comparative study focusing reasons for partial success of EEA in GF(7^3) is made in Sect. 2.5.

2.1 The EEA with Examples and Its Computational Algorithm

The Euclidean Algorithm, following notation, is stated as follows,

$$GCD(A, B) = C$$

where C is the **Greatest Common Divisor** of A and B, for $A \geq B$. It may be noted that $GCD(A, 0) = A$, since A divides both A and 0. The Euclidean Algorithm can thus be stated by the following recursive operation:

$GCD(A, B) = GCD(B, A \bmod B),$
Ex.1: $GCD(30, 12) = GCD(12, 6) = GCD(6, 0) = 6$
Ex.2: $GCD(15, 8) = GCD(8, 7) = GCD(7, 1) = GCD(1, 0) = 1$

The EEA finds the values of two integers S and T such that,

$$S \times A + T \times B = \mathrm{GCD}(A, B). \tag{1}$$

If one adopts EEA in the two examples cited above, one gets the values of $S = 1$ and $T = -2$ for Ex. 1 and $S = -1$ and $T = 2$ for Ex. 2 and can write the two examples as follows:

$$\mathrm{Ex.1:}(1) \times 30 + (-2) \times 12 = \mathrm{GCD}(30, 12) = 6.$$
$$\mathrm{Ex.2:}(-1) \times 15 + (2) \times 8 = \mathrm{GCD}(15, 8) = 1.$$

2.1.1 Computational Algorithm to Find S and T

The EEA finds $\mathrm{GCD}(A, B)$ and at the same time calculates the values of S and T. The algorithm of computation is as follows:

Initialization: $S_1 = 1$, $T_1 = 0$, $S_2 = 0$, $T_2 = 1$,

Operations for next iteration(s):

$$Q \leftarrow A/B, \quad R \leftarrow A \bmod B = A - Q \times B. \tag{2a}$$

$$R_s \leftarrow S_1 - Q \times S_2, R_t \leftarrow T_1 - Q \times T_2. \tag{2b}$$

$$(S_1, T_1, A) \leftarrow (S_2, T_2, B). \tag{2c}$$

$$(S_2, T_2, B) \leftarrow (R_s, R_t, R). \tag{2d}$$

The above operations are continued until $R = 0$. When $R = 0$, one considers $S = S_1$, $T = T_1$, and $\mathrm{GCD}(A, B) =$ upgraded (A). Table 1 represents the recursive algorithm in tabular form.

Here, it is to be noted that Q and R are, respectively, the quotient and remainder of the division of A by B at all stages, while R_s is the remainder of a virtual division of S_1 by S_2 with Q as the quotient and R_t is the same with T_1 and T_2. The recursive algorithm stated in Table 1 considers three arrays, namely Dividend array (DD) with three elements S_1, T_1, A; Divisor array (DR) with three elements S_2, T_2, B; and Division array (DN) with two elements Q and R. In the paper, the symbol $_nS_1$ is read as S_1 in nth iteration.

At the nth step of table 1, the values of S and T are those obtained in DD[0] and DD [1], respectively, and the $\mathrm{GCD}(A, B)$ is the value obtained in DD [2]. One finds that the expression $S \times A + T \times B = \mathrm{GCD}(A, B)$ is always true.

Table 1 Presentation of the Extended Euclidean Algorithm recursively computed

DN[]		DD[]			DR[]			Remark
DN[0]	DN[1]	DD[0]	DD[1]	DD[2]	DR[0]	DR[1]	DR[2]	
Q	R	S_1	T_1	A	S_2	T_2	B	#0
–	–	1	0	A	0	1	B	#1
$_1Q = A/B$	$_1R = A\%B$	$_1S_1 = 0$	$_1T_1 = 1$	$_1A = B$	$_1S_2$	$_1T_2$	$_1B = {}_1R$	#2
$_2Q = {}_1A/_1B$	$_2R = {}_1A\%_1B$	$_2S_1 = {}_1S_2$	$_2T_1 = {}_1T_2$	$_2A = {}_1B$	$_2S_2$	$_2T_2$	$_2B = {}_2R$	#3

The process is continued n-times till $_nR$ shown in DN[1] is zero

#0 variables location, #1 initialization, #2 Step 1: vide Eq. (2a, 2b, 2c, 2d), #3 Step 2: vide Eq. (2a, 2b, 2c, 2d)

2.2 Successful Application of EEA to Find Multiplicative Inverse Over GF(2^8)

If A and B are relatively prime polynomials, the GCD(A, B) is unity. From Eq. (1), one notes that

$$S \times A + T \times B = \text{GCD}(A,\ B) = 1$$
$$\text{Or} \qquad T \times B = 1 - S \times A$$

Hence, $T = B^{-1}$ mod A, since (T $\times$ B $-$ 1) is divisible by A.

It is desired to find multiplicative inverse of an elementary polynomial $B(x) = (x^7 + x^5 + x^3 + x)$ in GF(2^8) over the irreducible polynomial $A(x) = (x^8 + x^4 + x^3 + x + 1)$.

[0]	[1]	[2]
Initialization:		
DD[]: $S_1(x) = 1$;	$T_1(x) = 0$;	$A(x) = x^8 + x^4 + x^3 + x +1$
DR[]: $S_2(x) = 0$;	$T_2(x) = 1$;	$B(x) = x^7 + x^5 + x^3 + x$
Iteration 1:		
DN[]: $_1Q(x) = x$;	$_1R(x) = x^6 + x^3 + x^2 + x + 1$	
DD[]: $_1S_1(x) = 0$;	$_1T_1(x) = 1$;	$_1A(x) = x^7 + x^5 + x^3 + x$
DR[]: $_1S_2(x) = 1$;	$_1T_2(x) = x$;	$_1B(x) = x^6 + x^3 + x^2 + x + 1$
Iteration 2:		
DN[]: $_2Q(x) = x$;	$_2R(x) = x^5 + x^4 + x^2$	
DD[]: $_2S_1(x) = 1$;	$_2T_1(x) = x$;	$_2A(x) = x^6 + x^3 + x^2 + x + 1$
DR[]: $_2S_2(x) = x$;	$_2T_2(x) = x^2 + 1$;	$_2B(x) = x^5 + x^4 + x^2$
Iteration 3:		
DN[]: $_3Q(x) = x+1$;	$_3R(x) = x^4 + x+1$	
DD[]: $_3S_1(x) = x$;	$_3T_1(x) = x^2 + 1$;	$_3A(x) = x^5 + x^4 + x^2$
DR[]: $_3S_2(x) = x^2 + x +1$;	$_3T_2(x) = x^3 + x^2 + 1$;	$_3B(x) = x^4 + x + 1$
Iteration 4:		
DN[]: $_4Q(x) = x+1$;	$_4R(x) = 1$	
DD[]: $_4S_1(x) = x^2 + x + 1$;	$_4T_1(x) = x^3 + x^2 + 1$;	$_4A(x) = x^4 + x + 1$
DR[]: $_4S_2(x) = x^3 + x + 1$;	$_4T_2(x) = x^4 + x$;	$_4B(x) = 1$
Iteration 5:		
DN[]: $_5Q(x) = x^4 + x + 1$;	$_5R(x) = 0$	
DD[]: $_5S_1(x) = x^3 + x + 1$;	$_5T_1(x) = x^4 + x$;	$_5A(x) = 1$
DR[]: $_5S_2(x) = x^7 + x^5 + x^3 + x$;	$_5T_2(x) = x^8 + x^4 + x^3 + x + 1$;	$_5B(x) = 0$

In this example, GCD($A(x)$, $B(x)$) is $_5A(x) = 1$. Hence, the multiplicative inverse of $B(x)$ is $_5T_1(x)$ mod $A(x)$,

$$\left(x^7 + x^5 + x^3 + x\right)^{-1} = \left(x^4 + x\right).$$

2.3 Successful Application of EEA to Find Multiplicative Inverse Over GF(7^3)

There are 112 irreducible polynomials in GF(7^3), its list is available in [7, 8]. For IP, $I(x) = x^3 + 2x^2 + 6x + 1$ and EP $= 2x + 4$, corresponding to its value 18, the calculation of MI is shown below. Here, $I(x)$ is taken as $A(x)$ and EP as $B(x)$.

[0]	[1]	[2]
Initialization:		
DD[]: $S_1(x) = 1$;	$T_1(x) = 0$;	$A(x) = x^3 + 2x^2 + 6x + 1$
DR[]: $S_2(x) = 0$;	$T_2(x) = 1$;	$B(x) = 2x + 4$
Iteration 1:		
DN[]: $_1Q(x) = 4x^2 + 3$;	$_1R(x) = 3$	
DD[]: $_1S_1(x) = 0$;	$_1T_1(x) = 1$;	$_1A(x) = 2x + 4$
DR[]: $_1S_2(x) = 1$;	$_1T_2(x) = 3x^2 + 4$;	$_1B(x) = 3$
Iteration 2:		
DN[]: $_2Q(x) = 3x + 1$;	$_2R(x) = 1$	
DD[]: $_2S_1(x) = 1$;	$_2T_1(x) = 3x^2 + 4$	$_2A(x) = 3$
DR[]: $_2S_2(x) = 4x + 6$;	$_2T_2(x) = 5x^3 + 4x^2 + 2x + 4$;	$_2B(x) = 1$
Iteration 3:		
DN[]: $_3Q(x) = 3$;	$_3R(x) = 0$	
DD[]: $_3S_1(x) = 4x + 6$;	$_3T_1(x) = 5x^3 + 4x^2 + 2x + 4$;	$_3A(x) = 1$
DR[]: $_3S_2(x) = 2x + 4$;	$_3T_2(x) = 6x^3 + 5x^2 + x + 6$;	$_3B(x) = 0$

In the example, GCD($A(x)$, $B(x)$) is $_3A(x) = 1$ and multiplicative inverse of $B(x)$ is $_3T_1(x)$ mod $A(x)$,

$$(2x + 4)^{-1} = {_3T_1(x)} \bmod A(x)$$
$$= \left(5x^3 + 4x^2 + 2x + 4\right) \bmod \left(x^3 + 2x^2 + 6x + 1\right)$$
$$= x^2 + 6$$

Now to verify that $x^2 + 6$ is indeed multiplicative inverse of $2x + 4$ over GF(7^3) under irreducible polynomial $I(x) = x^3 + 2x^2 + 6x + 1$, we calculate the product of those two elements mod $I(x)$ to find 1,

$$(2x + 4)\left(x^2 + 6\right) \bmod I(x)$$
$$= \left(2x^3 + 4x^2 + 5x + 3\right) \bmod \left(x^3 + 2x^2 + 6x + 1\right)$$
$$= 1$$

2.4 Unsuccessful Application of EEA to Find Multiplicative Inverse Over GF(7^3)

The EEA is applied for a new $E(x) = 2x^2 + 5x + 3$ corresponding to its value 136. Here, $A(x) = I(x)$, $B(x) = E(x)$.

[0]	[1]	[2]
Initialization:		
DD[]: $S_1(x) = 1$;	$T_1(x) = 0$;	$A(x) = x^3 + 2x^2 + 6x + 1$
DR[]: $S_2(x) = 0$;	$T_2(x) = 1$;	$B(x) = 2x^2 + 5x + 3$
Iteration 1:		
DN[]: $_1Q(x) = 4x + 5$;	$_1R(x) = 4x$	
DD[]: $_1S_1(x) = 0$;	$_1T_1(x) = 1$;	$_1A(x) = 2x^2 + 5x + 3$
DR[]: $_1S_2(x) = 1$;	$_1T_2(x) = 3x + 2$;	$_1B(x) = 4x$
Iteration 2:		
DN[]: $_2Q(x) = 4x + 3$;	$_2R(x) = 3$	
DD[]: $_2S_1(x) = 1$;	$_2T_1(x) = 3x + 2$	$_2A(x) = 4x$
DR[]: $_2S_2(x) = 3x + 4$;	$_2T_2(x) = 2x^2 + 4x + 2$;	$_2B(x) = 3$
Iteration 3:		
DN[]: $_3Q(x) = 6x$;	$_3R(x) = 0$	
DD[]: $_3S_1(x) = 3x + 4$;	$_3T_1(x) = 2x^2 + 4x + 2$;	$_3A(x) = 3$
DR[]: $_3S_2(x) = 3x^2 + 4x + 1$;	$_3T_2(x) = 2x^3 + 4x^2 + 5x + 2$;	$_3B(x) = 0$

Here, $_3A(x) \neq 1$ when $_3R(x) = 0$, i.e., EEA fails to find GCD($A(x), B(x)$). It is observed that the EEA fails to find multiplicative inverse in this case.

2.5 Comparative Study of Success and Failures of EEA Over GF(2^8) and GF(7^3)

A complete list of successful and unsuccessful elemental polynomials under the $I(x) = x^3 + 2x^2 + 6x + 1$ corresponding to $(1261)_7$ is shown in Tables 2 and 3, respectively. In both the tables, the columns entitled "No.", "Polynomial", and "G" represents serial number, the elemental polynomial $E(x)$ and GCD($I(x), E(x)$), respectively.

The reason that EEA failed for an elemental polynomial $E(x)$ under an irreducible polynomial $I(x)$ over GF(7^3) is that the GCD($I(x), E(x)$) is not unity. It may be noted that the said GCD for a finite field GF(p^m) is supposed to vary between 1 and ($p - 1$). Table 2 lists 276 elemental polynomials, the GCD of each of all these polynomials with the irreducible polynomial $(1261)_7$ is observed to be unity. The multiplicative inverses of all these elemental polynomials can be successfully calculated following EEA under the irreducible polynomial $(1261)_7$. Table 3 lists 66 elemental polynomials the GCD of each of all these polynomials with the irreducible polynomial $(1261)_7$ is observed to non-unity varying between 2 and 6. However, for GF(2^8), there is no scope for the GCD being non-unity. It has been observed that the said GCD is unity for each of all the 30 irreducible polynomials with 255 elemental polynomials. In fact, EEA can be successfully applied for any kind of binary extension field.

3 Proposed Algebraic Method and Multiplicative Inverse Over GF(7^3)

We propose a new algebraic method that can successfully calculate multiplicative inverse of each element for all the 112 monic irreducible polynomials over GF(7^3). In Sect. 3.1, we discuss our proposed method and its application to calculate multiplicative inverses of the examples shown in Sects. 2.3 and 2.4 above, are presented in Sects. 3.2 and 3.3 respectively.

3.1 Algebraic Method to Find the Multiplicative Inverse Over GF(7^3)

Let $I(x) = (x^3 + a_2x^2 + a_1x + a_0)$ be a monic irreducible polynomial.

We want to find multiplicative inverse of $b(x) = (b_2x^2 + b_1x + b_0)$ under this irreducible polynomial.

If $c(x) = (c_2x^2 + c_1x + c_0)$ be MI, then we can write,

$$[b(x)\,c(x)] \bmod I(x) = 1$$

or, $$\left[(b_2x^2 + b_1x + b_0)\left(c_2x^2 + c_1x + c_0\right)\right] \bmod \left(x^3 + a_2x^2 + a_1x + a_0\right) = 1.$$

$$(3)$$

Here, the target is to find the values for c_2, c_1, and c_0. One can get these values by solving Eq. (3) as follows:

Table 2 List of elemental polynomials for which EEA is applied with success to find multiplicative inverse over irreducible polynomial $(1261)_7$

No.	Polynomial	G	No.	Polynomial	G	No.	Polynomial	G	No.	Polynomial	G
1	1	1	2	2	1	3	3	1	4	4	1
5	5	1	6	6	1	7	7	1	8	$x+1$	1
9	$x+2$	1	10	$x+3$	1	11	$x+4$	1	12	$x+5$	1
13	$2x$	1	14	$2x+1$	1	15	$2x+2$	1	16	$2x+3$	1
17	$2x+4$	1	18	$2x+5$	1	19	$2x+6$	1	20	$3x$	1
21	$3x+1$	1	22	$3x+2$	1	23	$3x+4$	1	24	$3x+5$	1
25	$4x$	1	26	$4x+1$	1	27	$4x+2$	1	28	$4x+3$	1
29	$4x+4$	1	30	$4x+5$	1	31	$4x+6$	1	32	$5x$	1
33	$5x+1$	1	34	$5x+2$	1	35	$5x+4$	1	36	$5x+5$	1
37	$5x+6$	1	38	$6x$	1	39	$6x+1$	1	40	$6x+2$	1
41	$6x+3$	1	42	$6x+4$	1	43	$6x+5$	1	44	$6x+6$	1
45	x^2	1	46	x^2+1	1	47	x^2+2	1	48	x^2+3	1
49	x^2+5	1	50	x^2+x	1	51	x^2+x+1	1	52	x^2+x+3	1
53	x^2+x+5	1	54	x^2+2x	1	55	x^2+2x+1	1	56	x^2+2x+2	1
57	x^2+2x+3	1	58	x^2+2x+4	1	59	x^2+2x+5	1	60	x^2+2x+6	1
61	x^2+3x	1	62	x^2+3x+1	1	63	x^2+3x+2	1	64	x^2+3x+3	1
65	x^2+3x+5	1	66	x^2+4x	1	67	x^2+4x+1	1	68	x^2+4x+5	1
69	x^2+5x	1	70	x^2+5x+1	1	71	x^2+5x+3	1	72	x^2+5x+5	1
73	x^2+5x+6	1	74	x^2+6x	1	75	x^2+6x+1	1	76	x^2+6x+3	1
77	x^2+6x+4	1	78	$2x^2$	1	79	$2x^2+1$	1	80	$2x^2+3$	1
81	$2x^2+5$	1	82	$2x^2+6$	1	83	$2x^2+x$	1	84	$2x^2+x+1$	1
85	$2x^2+x+2$	1	86	$2x^2+x+5$	1	87	$2x^2+2x$	1	88	$2x^2+2x+1$	1
89	$2x^2+2x+2$	1	90	$2x^2+2x+3$	1	91	$2x^2+2x+5$	1	92	$2x^2+2x+6$	1
93	$2x^2+3x$	1	94	$2x^2+3x+2$	1	95	$2x^2+3x+4$	1	96	$2x^2+3x+6$	1

(continued)

Table 2 (continued)

No.	Polynomial	G	No.	Polynomial	G	No.	Polynomial	G	No.	Polynomial	G
97	$2x^2 + 4x$	1	98	$2x^2 + 4x + 1$	1	99	$2x^2 + 4x + 2$	1	100	$2x^2 + 4x + 3$	1
101	$2x^2 + 4x + 4$	1	102	$2x^2 + 4x + 5$	1	103	$2x^2 + 5x$	1	104	$2x^2 + 5x + 1$	1
105	$2x^2 + 5x + 2$	1	106	$2x^2 + 5x + 5$	1	107	$2x^2 + 5x + 6$	1	108	$2x^2 + 6x$	1
109	$2x^2 + 6x + 1$	1	110	$2x^2 + 6x + 2$	1	111	$2x^2 + 6x + 3$	1	112	$2x^2 + 6x + 4$	1
113	$3x^2$	1	114	$3x^2 + 1$	1	115	$3x^2 + 2$	1	116	$3x^2 + 3$	1
117	$3x^2 + 4$	1	118	$3x^2 + 5$	1	119	$3x^2 + 6$	1	120	$3x^2 + x$	1
121	$3x^2 + x + 1$	1	122	$3x^2 + x + 2$	1	123	$3x^2 + x + 3$	1	124	$3x^2 + x + 4$	1
125	$3x^2 + x + 5$	1	126	$3x^2 + x + 6$	1	127	$3x^2 + 2x$	1	128	$3x^2 + 2x + 1$	1
129	$3x^2 + 2x + 4$	1	130	$3x^2 + 2x + 5$	1	131	$3x^2 + 3x$	1	132	$3x^2 + 3x + 1$	1
133	$3x^2 + 3x + 2$	1	134	$3x^2 + 3x + 3$	1	135	$3x^2 + 3x + 4$	1	136	$3x^2 + 3x + 6$	1
137	$3x^2 + 4x$	1	138	$3x^2 + 4x + 1$	1	139	$3x^2 + 4x + 2$	1	140	$3x^2 + 4x + 3$	1
141	$3x^2 + 4x + 4$	1	142	$3x^2 + 4x + 6$	1	143	$3x^2 + 5x$	1	144	$3x^2 + 5x + 1$	1
145	$3x^2 + 5x + 5$	1	146	$3x^2 + 6x$	1	147	$3x^2 + 6x + 1$	1	148	$3x^2 + 6x + 2$	1
149	$3x^2 + 6x + 4$	1	150	$3x^2 + 6x + 6$	1	151	$4x^2$	1	152	$4x^2 + 1$	1
153	$4x^2 + 2$	1	154	$4x^2 + 3$	1	155	$4x^2 + 4$	1	156	$4x^2 + 6$	1
157	$4x^2 + x$	1	158	$4x^2 + x + 1$	1	159	$4x^2 + x + 2$	1	160	$4x^2 + x + 3$	1
161	$4x^2 + x + 5$	1	162	$4x^2 + x + 6$	1	163	$4x^2 + 2x$	1	164	$4x^2 + 2x + 1$	1
165	$4x^2 + 2x + 4$	1	166	$4x^2 + 2x + 5$	1	167	$4x^2 + 3x$	1	168	$4x^2 + 3x + 1$	1
169	$4x^2 + 3x + 2$	1	170	$4x^2 + 3x + 3$	1	171	$4x^2 + 3x + 4$	1	172	$4x^2 + 3x + 5$	1
173	$4x^2 + 3x + 6$	1	174	$4x^2 + 4x + 1$	1	175	$4x^2 + 4x + 2$	1	176	$4x^2 + 4x + 3$	1
177	$4x^2 + 4x + 4$	1	178	$4x^2 + 4x + 5$	1	179	$4x^2 + 4x + 6$	1	180	$4x^2 + 5x$	1
181	$4x^2 + 5x + 1$	1	182	$4x^2 + 5x + 4$	1	183	$4x^2 + 5x + 5$	1	184	$4x^2 + 6x$	1
185	$4x^2 + 6x + 1$	1	186	$4x^2 + 6x + 2$	1	187	$4x^2 + 6x + 3$	1	188	$4x^2 + 6x + 4$	1

(continued)

Table 2 (continued)

No.	Polynomial	G	No.	Polynomial	G	No.	Polynomial	G	No.	Polynomial	G
189	$4x^2 + 6x + 5$	1	190	$4x^2 + 6x + 6$	1	191	$5x^2$	1	192	$5x^2 + 1$	1
193	$5x^2 + 2$	1	194	$5x^2 + 3$	1	195	$5x^2 + 4$	1	196	$5x^2 + 5$	1
197	$5x^2 + 6$	1	198	$5x^2 + x$	1	199	$5x^2 + x + 1$	1	200	$5x^2 + x + 2$	1
201	$5x^2 + 2x$	1	202	$5x^2 + 2x + 1$	1	203	$5x^2 + 2x + 2$	1	204	$5x^2 + 2x + 3$	1
205	$5x^2 + 2x + 4$	1	206	$5x^2 + 2x + 5$	1	207	$5x^2 + 3x$	1	208	$5x^2 + 3x + 1$	1
209	$5x^2 + 3x + 2$	1	210	$5x^2 + 3x + 3$	1	211	$5x^2 + 3x + 4$	1	212	$5x^2 + 3x + 5$	1
213	$5x^2 + 3x + 6$	1	214	$5x^2 + 4x$	1	215	$5x^2 + 4x + 1$	1	216	$5x^2 + 4x + 2$	1
217	$5x^2 + 4x + 3$	1	218	$5x^2 + 4x + 4$	1	219	$5x^2 + 4x + 5$	1	220	$5x^2 + 4x + 6$	1
221	$5x^2 + 5x$	1	222	$5x^2 + 5x + 1$	1	223	$5x^2 + 5x + 2$	1	224	$5x^2 + 5x + 3$	1
225	$5x^2 + 5x + 4$	1	226	$5x^2 + 5x + 5$	1	227	$5x^2 + 5x + 6$	1	228	$5x^2 + 6x$	1
229	$5x^2 + 6x + 1$	1	230	$5x^2 + 6x + 2$	1	231	$5x^2 + 6x + 4$	1	232	$5x^2 + 6x + 5$	1
233	$5x^2 + 6x + 6$	1	234	$6x^2$	1	235	$6x^2 + 1$	1	236	$6x^2 + 3$	1
237	$6x^2 + 4$	1	238	$6x^2 + 5$	1	239	$6x^2 + 6$	1	240	$6x^2 + x$	1
241	$6x^2 + x + 2$	1	242	$6x^2 + x + 3$	1	243	$6x^2 + x + 4$	1	244	$6x^2 + x + 5$	1
245	$6x^2 + x + 6$	1	246	$6x^2 + 2x$	1	247	$6x^2 + 2x + 1$	1	248	$6x^2 + 2x + 2$	1
249	$6x^2 + 2x + 3$	1	250	$6x^2 + 2x + 4$	1	251	$6x^2 + 2x + 5$	1	252	$6x^2 + 2x + 6$	1
253	$6x^2 + 3x$	1	254	$6x^2 + 3x + 2$	1	255	$6x^2 + 3x + 3$	1	256	$6x^2 + 3x + 4$	1
257	$6x^2 + 3x + 5$	1	258	$6x^2 + 3x + 6$	1	259	$6x^2 + 4x$	1	260	$6x^2 + 4x + 1$	1
261	$6x^2 + 4x + 2$	1	262	$6x^2 + 4x + 3$	1	263	$6x^2 + 4x + 4$	1	264	$6x^2 + 4x + 5$	1
265	$6x^2 + 4x + 6$	1	266	$6x^2 + 5x$	1	267	$6x^2 + 5x + 1$	1	268	$6x^2 + 5x + 2$	1
269	$6x^2 + 5x + 3$	1	270	$6x^2 + 5x + 4$	1	271	$6x^2 + 5x + 5$	1	272	$6x^2 + 6x$	1
273	$6x^2 + 6x + 1$	1	274	$6x^2 + 6x + 2$	1	275	$6x^2 + 6x + 5$	1	276	$6x^2 + 6x + 6$	1

Table 3 List of elemental polynomials for which EEA is applied with failure to find multiplicative inverse over irreducible polynomial $(1261)_7$

No.	Polynomial	G	No.	Polynomial	G	No.	Polynomial	G	No.	Polynomial	G
1	$x+6$	3	2	$3x+3$	3	3	$3x+6$	3	4	$5x+3$	3
5	x^2+4	4	6	x^2+6	3	7	x^2+x+2	2	8	x^2+x+4	2
9	x^2+x+6	5	10	x^2+3x+4	2	11	x^2+3x+6	6	12	x^2+4x+2	2
13	x^2+4x+3	3	14	x^2+4x+4	4	15	x^2+4x+6	2	16	x^2+5x+2	2
17	x^2+5x+4	2	18	x^2+6x+2	2	19	x^2+6x+5	5	20	x^2+6x+6	2
21	$2x^2+2$	6	22	$2x^2+4$	2	23	$2x^2+x+3$	2	24	$2x^2+x+4$	5
25	$2x^2+x+6$	2	26	$2x^2+2x+4$	2	27	$2x^2+3x+1$	3	28	$2x^2+3x+3$	3
29	$2x^2+3x+5$	5	30	$2x^2+4x+6$	5	31	$2x^2+5x+3$	3	32	$2x^2+5x+4$	2
33	$2x^2+6x+5$	5	34	$2x^2+6x+6$	5	35	$3x^2+2x+2$	2	36	$3x^2+2x+3$	2
37	$3x^2+2x+6$	3	38	$3x^2+3x+5$	3	39	$3x^2+4x+5$	2	40	$3x^2+5x+2$	2
41	$3x^2+5x+3$	3	42	$3x^2+5x+4$	3	43	$3x^2+5x+6$	2	44	$3x^2+6x+3$	3
45	$3x^2+6x+5$	5	46	$4x^2+5$	5	47	$4x^2+x+4$	3	48	$4x^2+2x+2$	2
49	$4x^2+2x+3$	2	50	$4x^2+2x+6$	2	51	$4x^2+4x$	6	52	$4x^2+5x+2$	2
53	$4x^2+5x+3$	3	54	$4x^2+5x+6$	3	55	$5x^2+x+3$	2	56	$5x^2+x+4$	3
57	$5x^2+x+5$	2	58	$5x^2+x+6$	2	59	$5x^2+2x+6$	2	60	$5x^2+6x+3$	2
61	$6x^2+2$	2	62	$6x^2+x+1$	2	63	$6x^2+3x+1$	2	64	$6x^2+5x+6$	6
65	$6x^2+6x+3$	2	66	$6x^2+6x+4$	2			5			

$$\left[b_2c_2x^4 + (b_1c_2 + b_2c_1)x^3 + (b_0c_2 + b_1c_1 + b_2c_0)x^2 + (b_0c_1 + b_1c_0)x + b_0c_0\right] \bmod \left(x^3 + a_2x^2 + a_1x + a_0\right) = 1$$

or, $\left[b_2c_2x(x^3 + a_2x^2 + a_1x + a_0) + (b_1c_2 + b_2c_1 - a_2b_2c_2)x^3 + (b_0c_2 + b_1c_1 + b_2c_0 - a_1b_2c_2)x^2 \right.$
$\left. + (b_0c_1 + b_1c_0 - a_0b_2c_2)x + b_0c_0\right] \bmod \left(x^3 + a_2x^2 + a_1x + a_0\right) = 1$

or, $\left[(b_1c_2 + b_2c_1 - a_2b_2c_2)\left(x^3 + a_2x^2 + a_1x + a_0\right)\right.$
$+ (b_0c_2 + b_1c_1 + b_2c_0 - a_1b_2c_2 - a_2b_1c_2 - a_2b_2c_1 + a_2^2b_2c_2)x^2$
$+ (b_0c_1 + b_1c_0 - a_0b_2c_2 - a_1b_1c_2 - a_1b_2c_1 + a_1a_2b_2c_2)x$
$\left. + (b_0c_0 - a_0b_1c_2 - a_0b_2c_1 + a_0a_2b_2c_2)\right] \bmod \left(x^3 + a_2x^2 + a_1x + a_0\right) = 1$

or, $\left[\{(a_2^2b_2 - a_1b_2 - a_2b_1 + b_0)c_2 + (b_1 - a_2b_2)c_1 + b_2c_0\}x^2 \right.$
$+ \{(a_1a_2b_2 - a_0b_2 - a_1b_1)c_2 + (b_0 - a_1b_2)c_1 + b_1c_0\}x$
$\left. + \{(a_0a_2b_2 - a_0b_1)c_2 - a_0b_2c_1 + b_0c_0\}\right] \bmod \left(x^3 + a_2x^2 + a_1x + a_0\right) = 1.$

$$(4)$$

From Eq. (4), it is evident that the dividend is smaller than the divisor. Hence, to satisfy the required condition of the remainder $= 1$, the following properties must hold.

1. The coefficients of $x^2 \equiv 0 \bmod 7$.
2. The coefficients of $x \equiv 0 \bmod 7$.
3. The constant part $\equiv 1 \bmod 7$.

Therefore,

$$\{(a_2^2b_2 - a_1b_2 - a_2b_1 + b_0)c_2 + (b_1 - a_2b_2)c_1 + b_2c_0\} \bmod 7 = 0. \qquad (5a)$$

$$\{(a_1a_2b_2 - a_0b_2 - a_1b_1)c_2 + (b_0 - a_1b_2)c_1 + b_1c_0\} \bmod 7 = 0. \qquad (5b)$$

$$\{(a_0a_2b_2 - a_0b_1)c_2 - a_0b_2c_1 + b_0c_0\} \bmod 7 = 1. \qquad (5c)$$

Note: Here, $GF(7^3)$ is used, and in modular arithmetic with modulus 7, the -1 is equivalent to $(-1 + 7) = 6$. Hence, the $-X$ in Eq. (5a, 5b, 5c) can be written as $+6X$. Accordingly, the Eq. (5a, 5b, 5c) becomes

$$\{(a_2^2b_2 + 6a_1b_2 + 6a_2b_1 + b_0)c_2 + (b_1 + 6a_2b_2)c_1 + b_2c_0\} \bmod 7 = 0. \quad (6a)$$

$$\{(a_1a_2b_2 + 6a_0b_2 + 6a_1b_1)c_2 + (b_0 + 6a_1b_2)c_1 + b_1c_0\} \bmod 7 = 0. \quad (6b)$$

$$\{(a_0a_2b_2 + 6a_0b_1)c_2 + 6a_0b_2c_1 + b_0c_0\} \bmod 7 = 1. \qquad (6c)$$

The above Eq. (6a, 6b, 6c) can be written as,

$$(k_{00}c_0 + k_{01}c_1 + k_{02}c_2) \bmod 7 = 0. \qquad (7a)$$

$$(k_{10}c_0 + k_{11}c_1 + k_{12}c_2) \bmod 7 = 0. \qquad (7b)$$

$$(k_{20}c_0 + k_{21}c_1 + k_{22}c_2) \bmod 7 = 1. \tag{7c}$$

where k-values are known and these are equal to,

$$k_{00} = (b_2)\,\%7 \quad k_{01} = (b_1 + 6a_2b_2)\,\%7 \quad k_{02} = \left(a_2^2 b_2 + 6a_1b_2 + 6a_2b_1 + b_0\right)\%7. \tag{8a}$$

$$k_{10} = (b_1)\,\%7 \quad k_{11} = (b_0 + 6a_1b_2)\,\%7 \quad k_{12} = (a_1a_2b_2 + 6a_0b_2 + 6a_1b_1)\,\%7. \tag{8b}$$

$$k_{20} = (b_0)\,\%7 \quad k_{21} = (6a_0b_2)\,\%7 \quad k_{22} = (a_0a_2b_2 + 6a_0b_1)\,\%7. \tag{8c}$$

The Eq. (7a, 7b, 7c), i.e., $(k \times c)\,\%7 = m$ can be solved by using matrix method as,

$$c = (k^{-1} \times m)\,\%7. \tag{9}$$

where

$$m = \begin{bmatrix} 0 \\ 0 \\ 1 \end{bmatrix}, k = \begin{bmatrix} k_{00} & k_{01} & k_{02} \\ k_{10} & k_{11} & k_{12} \\ k_{20} & k_{21} & k_{22} \end{bmatrix}, k^{-1} = \begin{bmatrix} ik_{00} & ik_{01} & ik_{02} \\ ik_{10} & ik_{11} & ik_{12} \\ ik_{20} & ik_{21} & ik_{22} \end{bmatrix}, c = \begin{bmatrix} c_0 \\ c_1 \\ c_2 \end{bmatrix}$$

$$= \begin{bmatrix} ik_{02} \\ ik_{12} \\ ik_{22} \end{bmatrix} \tag{10}$$

While calculating k^{-1} from k-matrix, one has to ensure that the determinant $\det(k)$ is nonzero. In the event $\det(k) = 0$, the $I(x)$ is not an irreducible polynomial, rather a reducible one and k^{-1} matrix for such a case does not exist. If $\det(k)$ is nonzero for all elements, the $I(x)$ is irreducible and the multiplicative inverses of elements exist. By calculating k^{-1} from k-matrix given in Eq. (10), one can get solution for c_0, c_1, and c_2 using Eq. (9).

Now, $(b_2x^2 + b_1x + b_0)^{-1} = (c_2x^2 + c_1x + c_0) \bmod (x^3 + a_2x^2 + a_1x + a_0)$

In the following Sects. 3.2 and 3.3, the two examples presented in Sects. 2.3 and 2.4 are successfully solved to provide the correct MI over GF(7^3).

3.2 Multiplicative Inverse of 2x + 4 in GF(7^3) by Using the Algebraic Method

Here, we would like to demonstrate the successful application of our proposed method to calculate the multiplicative inverse for a case shown in Sect. 2.3 where the EEA is successfully applied. The same irreducible and the elemental polynomials are used in the present case.

$$I(x) = x^3 + a_2x^2 + a_1x + a_0$$

Let the irreducible polynomial $I(x) = x^3 + a_2x^2 + a_1x + a_0$

$$= x^3 + 2x^2 + 6x + 1$$

The given polynomial $b(x) = b_2x^2 + b_1x + b_0$

$$= 2x + 4 \qquad (11)$$

One have to find $b(x)^{-1} = c(x) = c_2x^2 + c_1x + c_0.$

Here, $a_2 = 2, a_1 = 6, a_0 = 1$

$$b_2 = 0, b_1 = 2, b_0 = 4$$

By using these a and b values in Eq. (8a, 8b, 8c), one can calculate the k-values as,

$$
\begin{array}{lll}
k_{00} = 0\%7 = 0 & k_{01} = 2\%7 = 2 & k_{02} = 28\%7 = 0 \\
k_{10} = 2\%7 = 2 & k_{11} = 4\%7 = 4 & k_{12} = 72\%7 = 2 \\
k_{20} = 4\%7 = 4 & k_{21} = 0\%7 = 0 & k_{22} = 12\%7 = 5
\end{array}
$$

Following Eq. (10), the k-matrix and its inverse k^{-1} will be

$$
k = \begin{bmatrix} 0 & 2 & 0 \\ 2 & 4 & 2 \\ 4 & 0 & 5 \end{bmatrix}, \quad
k^{-1} = \begin{bmatrix} 2 & 6 & 6 \\ 4 & 0 & 0 \\ 4 & 5 & 1 \end{bmatrix}
$$

The solution of c in Eq. (9) will be obtained as $c_0, c_1,$ and c_2 from the last column of the k^{-1} matrix obtained above from k-matrix.

The solution for this problem is
$$
\begin{bmatrix} c_0 \\ c_1 \\ c_2 \end{bmatrix} = \begin{bmatrix} 6 \\ 0 \\ 1 \end{bmatrix}
$$

So one can obtain the required multiplicative inverse by using Eq. (11) as,

$$b(x)^{-1} = c(x)$$
$$= c_2x^2 + c_1x + c_0$$
$$= x^2 + 6$$

Hence, $(2x + 4)^{-1} = x^2 + 6$

The result is identical to that obtained in Sect. 2.3 above. The proposed method successfully finds the multiplicative inverse for this polynomial.

3.3 Multiplicative Inverse of $2x^2 + 5x + 3$ in GF(7^3) by Using the Algebraic Method

Now, we would like to demonstrate the successful application of our proposed method to calculate the multiplicative inverse for a case shown in Sect. 2.4 where the EEA has failed. The irreducible and the elemental polynomials used in the present case are the same used in Sect. 2.4.

$$\text{Let the irreducible polynomial } I(x) = x^3 + a_2x^2 + a_1x + a_0$$
$$= x^3 + 2x^2 + 6x + 1$$
$$\text{The given polynomial } b(x) = b_2x^2 + b_1x + b_0$$
$$= 2x^2 + 5x + 3 \tag{12}$$
$$\text{One have to find } b(x)^{-1} = c(x) = c_2x^2 + c_1x + c_0.$$
$$\text{Here, } a_2 = 2, a_1 = 6, a_0 = 1$$
$$b_2 = 2, b_1 = 5, b_0 = 3$$

By using these a and b values in Eq. (8a, 8b, 8c), one can calculate the k-values as,

$$
\begin{array}{lll}
k_{00} = 2\%7 = 2 & k_{01} = 29\%7 = 1 & k_{02} = 143\%7 = 3 \\
k_{10} = 5\%7 = 5 & k_{11} = 75\%7 = 5 & k_{12} = 216\%7 = 6 \\
k_{20} = 3\%7 = 3 & k_{21} = 12\%7 = 5 & k_{22} = 34\%7 = 6
\end{array}
$$

Following Eq. (10), the k-matrix and its inverse k^{-1} will be

$$
k = \begin{bmatrix} 2 & 1 & 3 \\ 5 & 5 & 6 \\ 3 & 5 & 6 \end{bmatrix}, \quad
k^{-1} = \begin{bmatrix} 0 & 4 & 3 \\ 4 & 6 & 6 \\ 6 & 0 & 3 \end{bmatrix}
$$

The solution of c in Eq. (9) will be obtained as c_0, c_1, and c_2 from k^{-1} matrix in Eq. (10).

The solution for this problem is $\begin{bmatrix} c_0 \\ c_1 \\ c_2 \end{bmatrix} = \begin{bmatrix} 3 \\ 6 \\ 3 \end{bmatrix}$

So one can obtain the required multiplicative inverse by using Eq. (12) as,

$$b(x)^{-1} = c(x)$$
$$= c_2x^2 + c_1x + c_0$$
$$= 3x^2 + 6x + 3$$
$$\text{Hence, } (2x^2 + 5x + 3)^{-1} = 3x^2 + 6x + 3$$

Now to verify that $(3x^2 + 6x + 3)$ is indeed the MI of $(2x^2 + 5x + 3)$ over $GF(7^3)$ under irreducible polynomial $I(x) = (x^3 + 2x^2 + 6x + 1)$, we calculate the product of these two elements mod $I(x)$ to find 1, that is,

$$\left(2x^2 + 5x + 3\right)\left(3x^2 + 6x + 3\right) \bmod I(x)$$
$$= \left(6x^4 + 6x^3 + 3x^2 + 5x + 2\right) \bmod \left(x^3 + 2x^2 + 6x + 1\right)$$
$$= 1.$$

The result is indeed correct.

4 Computational Algorithm

The strength of the proposed algebraic method is that it finds the first monic irreducible polynomial and then all the 342 multiplicative inverses under it. The algorithm continues computation till the last monic irreducible polynomial is obtained along with all the 342 multiplicative inverses under it. It is noticed that in course of computation, the algorithm finds all the 112 monic irreducible polynomials reported in [7, 8]. For all the monic irreducible polynomials over $GF(7^3)$, the coefficient of the x^3-term is taken as unity and for computing other coefficients of such polynomials, one needs to vary the loop-index from 343 to 685 corresponding to septenary equivalents of $(1000)_7$ and $(1666)_7$, respectively.

An indigenous C program entitled "GF7^3INV," consisting of an EP-loop with loop-index (ep) varying from 1 to 342 within an IP-loop with loop-index (ip) varying from 343 to 685 and two subprograms, is developed. In the IP-loop, the IP-coefficients are calculated based on IP-loop-index (ip) using coeff_pol(), stored in array a[] and then the cal_inverse() is called after entering the EP-loop. The cal_inverse() first calls the coeff_pol() to calculate EP-coefficients based on loop-index (ep) and to store them in array b[] and then using the arrays a[] and b[] values, the k-matrix, given in Eq. (10), is formed and the determinant det(k) is calculated. If det(k) = 0, it concludes that the current ip is not an irreducible polynomial and takes the next ip-index for subsequent computation. If det(k) $\neq$ 0, the k^{-1} is calculated whose third column is the array c[] shown in Eq. (10). Program algorithm for GF7^3INV is described below in pseudo-code:

```
Step 1: For ip = 343 to 685 do the following steps.
Step 2: Convert the ip into its septenary equivalent and
        store them in an array a[] defined in Eq.(3) where
        a₀ is the least significant septenary digit.
Step 3: For ep = 1 to 342 do the following steps.
Step 4: Convert the ep into its septenary equivalent and
        store them in an array b[] defined in Eq.(3) where b₀
        is the least significant septenary digit.
```

```
Step 5: From arrays a[] and b[] form the 3 × 3 k-matrix
        described in Eq. (10).
Step 6: Calculate determinant of k-matrix det(k).
Step 7: If det(k) = 0, go to step 10, otherwise find the
        inverse of k-matrix as k⁻¹-matrix.
Step 8: The result for the c coefficients in Eq. (3) is
        obtained from Eq. (10) as c₀=ik₀₂, c₁=ik₁₂, c₂=ik₂₂
Step 9: Go to step 3 for next ep.
Step 10: Go to step 1 for next ip.
Step 11: Stop.
```

The above algorithm generates a full list of inverses corresponding to all the 112 monic irreducible polynomials given in [7, 8] and confirms that no other monic irreducible polynomial exists. Under a particular monic polynomial, if the corresponding determinant $\det(k) \neq 0$ for a particular elemental polynomial, the algorithm calculates its multiplicative inverse and if $\det(k) = 0$, the algorithm stops looking for further multiplicative inverses and declares that the current monic polynomial is not an irreducible one.

5 Results, Discussion, and Future Scopes

A list of multiplicative inverses of 342 elemental polynomials from 1 to 342 for two irreducible polynomials ($x^3 + 2$) and ($x^3 + 2x^2 + 6x + 1$) over GF(7^3) is, respectively, given in two blocks of Table 4. In each block, the multiplicative inverses of all elemental polynomials are given sequentially. In 3rd block of Table 4, first 7 multiplicative inverses for a polynomial ($x^3 + 1$), supposed to an irreducible one, are shown. For eighth elemental polynomial, no multiplicative inverse exists indicating that the polynomial supposed to be irreducible is a reducible. Detailed results of multiplicative inverses for all the irreducible polynomials over GF(7^3) are given in a file "Result—MIs over GF7^3.pdf" entitled "Multiplicative Inverses of all the 342 Elemental Polynomials (EP) for all the 112 Irreducible Polynomials (IP) Over GF(7^3)" [10].

For future application scope of the proposed method in the field of cryptography, it is foreseen that a 9-bit pseudo-random number generator derived from the proposed method can be compared with an identical generator based on GF(2^9). Similarly, one can generate 8-bit pseudo-random number generator based on GF(7^3) and compare it with an identical generator based on GF(2^8). There is a future scope of mathematical interest of the proposed method. It is now possible to find the list of non-monic polynomials over GF(7^3). To the best knowledge of the authors, no such information is available in literature. The approach presented here to study monic and non-monic irreducible polynomials can be extended to higher-order prime extension fields.

Table 4 A sequential list of MIs of 342 EPs for IPs $(x^3 + 2)$ and $(x^3 + 2x^2 + 6x + 1)$ over GF(7^3)

(1) Multiplicative inverses of 342 EPs under IP $(x^3 + 2)$ corresponding to ip-index 345:

001, 004, 005, 002, 003, 006, 300, 616, 623, 214, 645, 241, 222, 500, 326, 343, 124, 315, 111, 142, 100, 365, 356, 252, 333, 264, 231, 600, 546, 513, 444, 525, 421, 412, 200, 635, 666, 462, 653, 434, 451, 400, 555, 536, 132, 563, 154, 161, 030, 413, 115, 643, 216, 346, 545, 461, 025, 636, 165, 363, 102, 135, 262, 335, 566, 433, 023, 463, 404, 521, 240, 063, 215, 604, 116, 310, 164, 665, 026, 201, 533, 236, 266, 324, 502, 440, 213, 065, 610, 416, 622, 066, 301, 510, 140, 113, 415, 050, 143, 245, 523, 446, 626, 325, 432, 655, 336, 153, 013, 133, 104, 234, 535, 016, 401, 353, 456, 436, 512, 036, 601, 340, 220, 243, 145, 131, 015, 556, 235, 633, 202, 255, 311, 420, 033, 445, 504, 246, 640, 614, 302, 120, 443, 035, 540, 146, 010, 162, 261, 651, 464, 352, 554, 136, 250, 351, 403, 652, 024, 660, 435, 552, 354, 360, 150, 206, 021, 625, 503, 524, 034, 514, 121, 212, 233, 551, 450, 022, 654, 560, 105, 526, 312, 305, 224, 322, 411, 032, 323, 621, 611, 114, 606, 031, 422, 060, 223, 425, 313, 126, 516, 615, 454, 355, 046, 101, 663, 166, 156, 251, 045, 366, 455, 553, 402, 465, 544, 602, 210, 123, 055, 320, 226, 152, 565, 656, 263, 043, 253, 204, 342, 056, 501, 620, 410, 423, 225, 641, 110, 053, 125, 304, 426, 520, 020, 452, 151, 331, 254, 532, 634, 163, 631, 230, 042, 334, 650, 405, 466, 130, 531, 203, 332, 044, 350, 646, 522, 505, 144, 542, 221, 062, 265, 632, 534, 550, 430, 106, 041, 543, 341, 321, 424, 306, 061, 242, 345, 603, 644, 064, 624, 441, 122, 040, 232, 431, 561, 134, 662, 364, 155, 362, 664, 630, 260, 406, 011, 453, 361, 160, 012, 564, 330, 205, 613, 511, 541, 244, 506, 051, 112, 256, 460, 661, 103, 562, 014, 530, 515, 303, 314, 054, 344, 211, 442, 316, 642, 605, 414, 612, 141, 052

(2) Multiplicative inverses of 342 EPs under IP $(x^3 + 2x^2 + 6x + 1)$ corresponding to ip-index 484:

001, 004, 005, 002, 003, 006, 651, 223, 205, 346, 620, 630, 264, 364, 310, 115, 350, 106, 132, 523, 245, 210, 162, 331, 326, 230, 304, 532, 403, 540, 451, 446, 615, 560, 413, 254, 645, 601, 420, 662, 460, 126, 513, 140, 150, 431, 502, 554, 646, 154, 336, 164, 122, 426, 024, 435, 243, 306, 516, 466, 022, 421, 265, 612, 104, 352, 543, 222, 060, 153, 600, 025, 255, 551, 215, 240, 062, 621, 443, 362, 525, 406, 263, 063, 340, 536, 130, 101, 442, 636, 313, 566, 032, 440, 103, 656, 266, 323, 411, 462, 213, 553, 012, 432, 031, 616, 314, 203, 225, 135, 531, 256, 233, 125, 011, 503, 214, 643, 035, 404, 520, 221, 653, 353, 450, 136, 625, 341, 111, 402, 030, 561, 545, 405, 633, 363, 051, 133, 220, 465, 664, 300, 146, 016, 120, 166, 262, 632, 526, 546, 036, 533, 112, 021, 663, 510, 160, 212, 505, 412, 541, 354, 344, 200, 360, 445, 034, 614, 033, 452, 361, 635, 622, 102, 151, 242, 660, 422, 322, 501, 013, 023, 602, 123, 235, 321, 434, 661, 324, 333, 143, 253, 020, 461, 506, 515, 665, 244, 041, 231, 251, 145, 453, 201, 316, 050, 524, 634, 444, 054, 116, 343, 456, 542, 654, 105, 626, 064, 206, 455, 355, 110, 535, 163, 605, 155, 142, 416, 325, 044, 236, 043, 332, 410, 500, 433, 423, 056, 365, 202, 565, 610, 260, 114, 454, 345, 065, 224, 564, 315, 366, 312, 611, 650, 061, 631, 400, 113, 232, 550, 644, 026, 414, 144, 302, 641, 216, 040, 305, 666, 436, 152, 042, 320, 424, 124, 556, 250, 303, 521, 134, 563, 204, 066, 652, 544, 046, 246, 642, 552, 504, 463, 161, 131, 053, 351, 655, 613, 441, 623, 464, 511, 121, 604, 330, 045, 211, 014, 141, 335, 606, 640, 241, 430, 015, 514, 301, 252, 415, 334, 156, 624, 530, 562, 226, 522, 052, 100, 512, 010, 555, 234, 425, 603, 165, 342, 356, 055, 311, 261, 401, 534

(3) Multiplicative inverses of first 7 EPs under IP $(x^3 + 1)$ corresponding to ip-index 344:

001, 004, 005, 002, 003, 006, 600, No inverse exists for 8th EP $(x + 1)$

6 Conclusion

A simple algebraic method is proposed in the paper to mark all possible monic irreducible polynomials over GF(7^3) followed by calculating multiplicative inverses of 342 elemental polynomials under each of all the 112 irreducible polynomials. Using the proposed method, one can have a mathematical look toward monic as well as non-monic irreducible polynomials over GF(p^m) for necessary values of p and m. The integers over GF(7^3) require nine bits—the same is also required for integers over GF(2^9). With array of multiplicative inverses of both types of finite fields, one would be able to form two types of 9-bit random number generators, both of which can be used in stream cipher and in block cipher. From the quantitative measure of randomness of the two, one would able to conclude which of the two is better and what is the status of those in respect of similar type of 8-bit generator.

Acknowledgments We express our gratitude toward UGC, New Delhi, for providing financial support to the first author. We are also indeed thankful to the Head of the Department of Radio Physics and Electronics, University of Calcutta, for providing necessary infrastructural facilities to undertake research activities.

References

1. Stallings, W.: Cryptography and Network Security Principles and Practices, 4th edn. Pearson Education, Delhi (2008)
2. Forouzan, B.A., Mukhopadhyay, D.: Cryptography and Network Security, 2nd edn. TMH, New Delhi (2011)
3. Arguello, F.: Lehmer-based algorithm for computing inverses in Galois fields GF(2^m). Electron. Lett. IET J. Mag. **42**(5), 270–271 (2006)
4. Hasan, M.A.: Double-basis multiplicative inversion over GF(2^m). IEEE Trans. Comput. **47**(9), 960–970 (1998)
5. Daemen, J., Rijmen, V.: AES Proposal: Rijndael, AES Algorithm Submission, 3 Sept 1999
6. FIPS Pub. 197, Announcing the Advanced Encryption Standard (AES), 26 Nov 2001
7. Church, R.: Tables of irreducible polynomials for the first four prime moduli. Ann. Math. **36**(1), 198–209 (1935)
8. Lidl, R., Niederreiter, H.: Finite fields, encyclopedia of mathematics and its applications, vol. 20. Addison-Wesley Publishing Company, Boston (1983)
9. Knuth, D.E.: The Art of Computer Programming Seminumerical Algorithms, 3rd edn, vol. 2. Pearson Education, Upper Saddle River (2011)
10. https://www.academia.edu/attachments/32975915/download_file

A Novel Biometric Template Encryption Scheme Using Sudoku Puzzle

Arnab Kumar Maji and Rajat Kumar Pal

Abstract Identity theft is a growing concern in the digital era. As per the US Federal Trade Commission, millions of people got victimized in each year [1]. Traditional authentication methods such as passwords and identity documents are not sufficient to combat ID theft or ensure security. Such representations of identity can easily be forgotten, lost, guessed, stolen, or shared. On the contrary, biometric systems recognize individuals based on their anatomical traits (e.g., fingerprint, face, palm print, iris, and voice) or behavioral traits (e.g., signature, gait). As such traits are physically linked to the user, biometric recognition is a natural and more reliable mechanism for ensuring that only legitimate or authorized users are able to enter a facility, access a computer system, or cross international borders. Biometric systems also offer unique advantages such as deterrence against repudiation and the ability to detect whether an individual has multiple identity cards (e.g., passports) under different names. Thus, biometric systems impart higher levels of security when appropriately integrated into applications requiring user authentication. In this paper, an attempt has been made to secure the biometric data using sudoku puzzle.

Keywords Biometric · Trait · Encryption · Decryption · Sudoku · Key · Minigrid · Backtracking

A.K. Maji (✉)
Department of Information Technology, North Eastern Hill University,
Shillong 793 022, India
e-mail: arnab.maji@gmail.com

R.K. Pal
Department of Computer Science and Engineering, University of Calcutta,
Kolkata 700 009, India
e-mail: pal.rajatk@gmail.com

© Springer India 2015   109
R. Chaki et al. (eds.), *Applied Computation and Security Systems*, Advances in Intelligent
Systems and Computing 305, DOI 10.1007/978-81-322-1988-0_7

1 Introduction

Biometric authentication schemes have great potentials in building secured systems since biometric data of the users are bound tightly with their identities and cannot be forgotten. Typically, a biometric authentication scheme consists of two phases, i.e., enrollment phase and authentication phase [2]. During the enrollment phase, the sensor module acquires the raw biometric data of an individual in the form of an image, video, audio, or some other signal. The feature extraction module operates on the biometric signal and extracts a salient set of features to represent the signal; during user enrollment, the extracted feature set, labeled with the user's identity, is stored in the biometric system and is known as a template. This template is stored on some central server or a device.

During the authentication phase, user would provide another biometric sample to the sensor. Features extracted from this sample constitute the query, which the system then compares to the template of the claimed identity via a biometric matcher. The matcher returns a match score representing the degree of similarity between the template and the query. The system accepts the identity claim only if the match score is above a predefined threshold. The whole scheme is presented in Fig. 1, where X and Y are templates.

Now, we briefly explain the activities performed in Fig. 1. During the enroll-ment phase of biometric authentication, the images of the traits are captured using some camera or devices. Then, features such as characteristics of the captured image such as color, pattern of the image, are extracted. Then, a biometric tem-plate (say, X) is generated from it. This template contains the characteristics of the user's biometric information that could be used to identify the said trait uniquely. Further, a sketch is generated from the template (i.e., X). A sketch is essentially a graph-like structure from which the template can be reconstructed, as and when necessary. The generated sketch is stored in the database. This part follows the upper row of the figure up to database. Now during the authentication phase, as

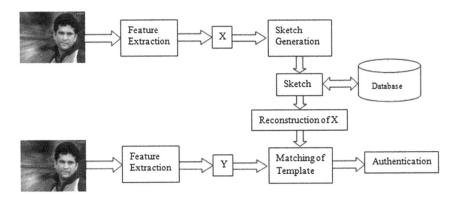

Fig. 1 Sketch generation and template reconstruction of biometric information

usual at first the trait's image is captured; then in the same way, features are extracted from it, and another template (say, Y) is constructed. Then, the template X is reconstructed from the sketch stored in the database. Afterward it is matched with the captured template (i.e., Y) by using some matching function. If the templates match, then we may say that the user is authenticated; otherwise, not.

Thus, a biometric system may be viewed as a pattern recognition system whose function is to classify a biometric signal into one of several identities (viz., identification) or into one of two classes—genuine and impostor users (viz., verification).

While a biometric system can enhance user convenience and security, it is also susceptible to various types of threats [3, 4] as discussed below in the next section.

1.1 Biometric System Vulnerabilities

A biometric system is vulnerable to two types of failures. A denial of service occurs when the system does not recognize a legitimate user, while an intrusion refers to the scenario in which the system incorrectly identifies an impostor as an authorized user. While there are many possible reasons for these failures, they can broadly be categorized as intrinsic limitations and adversary attacks.

1.1.1 Intrinsic Limitations

Unlike a password-based authentication system, which requires a perfect match between two alphanumeric strings, a biometric-based authentication system relies on the similarity between two biometric samples. This is because an individual's biometric sample acquired during enrollment and authentication is seldom identical; a biometric system can make two types of authentication errors. A false nonmatch occurs when two samples from the same individual have low similarity and the system cannot correctly match them. A false match occurs when two samples from different individuals have high similarity and the system incorrectly declares them as a match. A false nonmatch leads to a denial of service to a legitimate user, while a false match can result in intrusion by an impostor. This is because the impostor need not exert any special effort to fool the system; such an intrusion is known as a zero-effort attack. Most of the research endeavor in the biometrics community over the past five decades has focused on improving authentication accuracy—that is, on minimizing false nonmatches and false matches.

1.1.2 Adversary Attacks

A biometric system may also fail to operate as intended due to manipulation by adversaries. Such manipulations can be carried out via insiders, such as system administrators, or by directly attacking the system infrastructure. An adversary can circumvent a biometric system by coercing or colluding with insiders, exploiting their negligence (for example, failure to properly log out of a system after completing a transaction), or fraudulently manipulating the procedures of enrollment and exception processing, originally designed to help authorized users.

External adversaries can also cause a biometric system to fail through direct attacks on the user interface (sensor), the feature extractor and matcher modules, the interconnections between the modules, and the template database.

Examples of attacks targeting the system modules and their interconnections include Trojan horse, man-in-the-middle, and replay attacks. As most of these attacks are also applicable to password-based authentication systems, several countermeasures such as cryptography, time stamps, and mutual authentication are available to prevent them or minimize their impact. Two major vulnerabilities specifically deserve attention in the context of biometric authentication and ID cards, and it is not possible to replace stolen templates with new ones because biometric traits are irrevocable. Finally, the stolen biometric templates can be used for unintended purposes—for example, to covertly track a person across multiple systems or obtain private health information.

There are several possible reasons for these attacks. One of the most common reasons behind this type of attack is stealing of templates and modifying them [5]. So securing the biometric template is immensely important. At the same time, the quality of the template should not be degraded. In our proposed scheme, we have used the puzzle of a 9×9 sudoku instance as key to encrypt the biometric template. The novelty of our proposed scheme lies on encryption of the template with less distortion as well as it is expected to be incredibly difficult for the intruder to know about the keys and the intruder is not be able to change or distort the template.

2 Introduction to Sudoku

'Sudoku' is a popular Japanese puzzle game. It is usually a 9×9 grid-based puzzle problem which is subdivided into nine 3×3 minigrids, wherein some clues are given and the objective of the problem is to fill it up for the remaining blank positions. Furthermore, the objective of this problem is to compute a solution where the numbers 1 through 9 occur exactly once in each row, exactly once in each column, and exactly once in each minigrid independently obeying the given clues. An instance of a sudoku puzzle with its solution is shown in Fig. 2. Besides the standard 9×9 grid, variants of sudoku puzzles include some of the following.

- 4×4 grid with four 2×2 minigrids,
- 5×5 grid with pentomino regions published under the name Logi-5 [6]; a pentomino is composed of five congruent squares, connected orthogonally; pentomino is seen in playing the game Tetris [7],

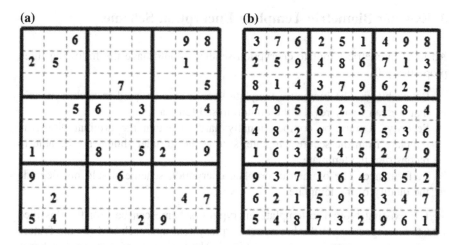

Fig. 2 a An instance of the sudoku problem. **b** A solution of the sudoku instance shown in **a** where a digit /symbol occurs exactly once in each row, column, and minigrid

- 8 × 8 grid with eight 2 × 4 minigrids [7],
- 16 × 16 grid (super sudoku) with 16 4 × 4 minigrids [8],
- 25 × 25 grid (sudoku, the Giant) with 25 5 × 5 minigrids [9], etc.

There are several logical techniques to solve the sudoku puzzle; some are basic simple logic, some are more advanced [8]. Depending on the difficulty of the puzzle [9], a blend of techniques may be needed in order to solve a puzzle. In fact, most computer-generated sudoku puzzles rank the difficulty based upon the number of empty cells in the puzzle and how much effort is needed to solve each of them. In our proposed scheme, we have used the minigrid backtracking [8] method to solve a sudoku puzzle.

The basic backtracking algorithm works as follows. The program places number 1 in the first empty cell. If the choice is compatible with the existing clues, it continues to the second empty cell, where it places a 1 (in some other row, column, and minigrid). When it encounters a conflict (which can happen very quickly), it erases the 1 a moment ago placed and inserts 2 or, if that is invalid, 3 or the next legal number. After placing the first legal number possible, it moves to the next cell and starts again with a 1 (or a minimum possible acceptable value). If the number that has to be altered is a 9, which cannot be raised by one in a standard 9 × 9 sudoku grid, the process backtracks and increases the number in the previous cell (or the next to the last number placed) by one. Then, it moves forward until it hits a new conflict. In this way, the process may sometimes backtrack several times before advancing. It is guaranteed to find a solution if there is one, simply because it eventually tries every possible number in every possible location.

3 Existing Biometric Template Encryption Scheme

An ideal biometric template encryption scheme should have the following three properties [10]:

Revocability: The biometric template should be encrypted in such a manner that we can easily revoke the compromised template.

Security: It must be computationally hard to obtain the original biometric template from the secure template, so that the intruder should not be able to reconstruct the template.

Performance: The biometric template encryption scheme should not degrade the quality of the template.

The existing biometric template encryption scheme can be broadly classified into two categories: (1) the biometric cryptosystem approach and (2) the feature transformation approach as shown in Fig. 3. The basic idea of these approaches is that instead of storing the original template, the transformed/encrypted template which is intended to be more secure is stored. In case the transformed/encrypted template is stolen or lost, it is computationally hard to reconstruct the original template and to determine the original raw biometric data simply from the transformed/encrypted template.

In the feature transformation approach, a transformation function (F) is applied to the biometric template (T) and only the transformed template $(F(T, K))$ is stored in the database. The parameter of the transformation function is typically derived from a Random Key (K) or password. The same transformation function is applied to query features (Q), and the transformed query $(F(Q, K))$ is directly matched against the transformed template $(F(T, K))$. Depending on the characteristics of the transformation function F, the feature transformation schemes can be further categorized as salting and noninvertible transforms. In salting, F in invertible, that

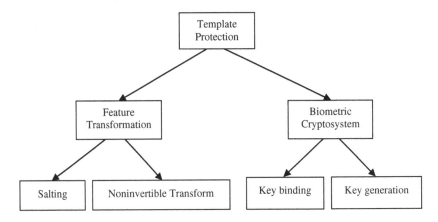

Fig. 3 Categorization of biometric template protection scheme

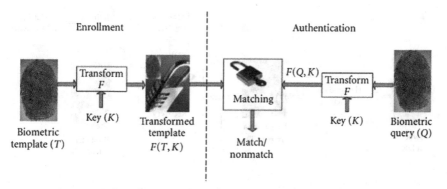

Fig. 4 Authentication mechanism when the biometric template is protected using feature transformation approach (courtesy to [5])

is if an adversary gains access to the key and the transformed template, s/he can recover the original biometric template (or a close approximation of it). Hence, the security of the salting scheme is based on the secrecy of the key or password. On the other hand, noninvertible transformation schemes typically apply a one-way function on the template and it is computationally hard to invert a transformed template even if the key is known. Figure 4 depicts the approach as described.

Biometric cryptosystems [11, 12] were originally developed for the purpose of either securing a cryptographic key using biometric features or directly generating a cryptographic key from biometric features. However, they can also be used as a template protection mechanism. In a biometric cryptosystem, some public information about the biometric template is stored. This public information is usually referred to as helper data, and hence, biometric cryptosystems are also known as helper data-based methods [13]. While the helper data do not (is not supposed to) reveal any significant information about the original biometric template, it is needed during matching to extract a cryptographic key from the query biometric features. Matching is performed indirectly by verifying the validity of the extracted key. Error correction coding techniques are typically used to handle intruder variations. Figure 5 depicts the approach as described above.

Biometric cryptosystems can further be classified as key-binding and key generation systems depending on how the helper data are obtained. When the helper data are obtained by binding a key (that is independent of the biometric features) with the biometric template, we refer to it as a key-binding biometric cryptosystem. Note that given only the helper data, it is computationally hard to recover either the key or the original template. Matching in a key-binding system involves recovery of the key from the helper data using the query biometric features. If the helper data are derived only from the biometric template and the cryptographic key is directly generated from the helper data and the query biometric features, it leads to a key generation biometric cryptosystem.

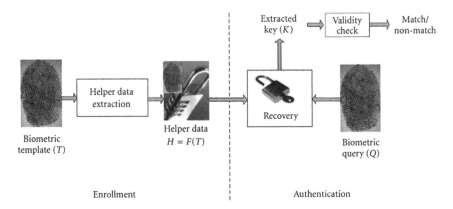

Fig. 5 Authentication mechanism using biometric cryptosystem (courtesy to [5])

4 Biometric Template Encryption Scheme Using Sudoku

The entire existing template encryption scheme can provide adequate security to the biometric template. But unfortunately, if somebody modifies the stored template, they are unable to detect it. So in our proposed scheme, we have used sudoku puzzle as a key to encrypt the template, so that if any modification in the template takes place, it can easily detect it. The entire scheme is described as below:

Input: A solved sudoku puzzle and the biometric template of a trait (e.g., an image).

Output: Sudoku-embedded biometric template.

Step 1 *Block preparation*: The biometric template is divided into 9×9 equal-sized blocks.

Step 2 *Embedding* 9×9 *sudoku puzzle*: A solved sudoku instance and the biometric template with blocks are taken as input.

For each individual block:
Make disjoint groups of four pixels each; padding is incorporated, if necessary.
For each group of four pixels:
The least significant bit (LSB) of the 8-bit representation of each pixel in block i is added to the associated value present in the corresponding sudoku cell i, $1 \leq i \leq 81$.

As for example, in Fig. 6, all pixels present in the first block of the first row are to be replaced with 3 in a group of four pixels each, i.e., the LSBs of the first and the second pixel are kept unchanged, whereas 1 is added to the LSB of each of the third and the fourth pixel, as the corresponding binary representation of 3 is 0011.

Step 3 *Sudoku instance and key generation*: Here, a suitable sudoku instance is generated (by excluding digits from some of the cells and making them blank) from a given solved sudoku puzzle. Hence, a sudoku instance is

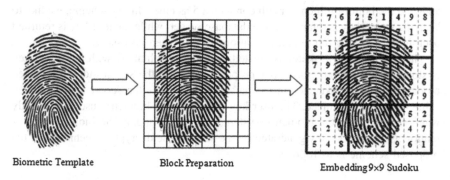

Biometric Template   Block Preparation   Embedding 9×9 Sudoku

Fig. 6 Biometric template encryption process: The biometric template is divided and placed over a region of 9 × 9 blocks. The values of each cell of a (solved) sudoku puzzle are embedded into the corresponding block of the template. Here, *red digits* are given clues of the problem instance and *black digits* are inserted to get a solved solution of the instance

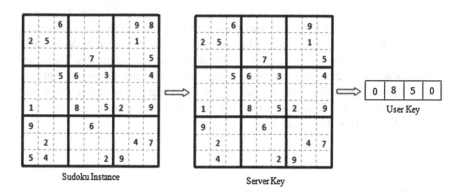

Fig. 7 The process of key generation

generated by keeping some of the digits as clues and rest of the cells remain blank so that at the end of the process the same solved sudoku puzzle could be obtained. This process is known as the *Dig-and-Hole* method for generating sudoku instance [9]. In Fig. 7, the clues are shown in red colors that are generated (in obtaining an instance of the problem) for some solved sudoku puzzle, as shown in Fig. 6, though this could also be a different instance other than that is depicted in this figure.

Then, two keys are created from this sudoku instance: (1) server key and (2) user key. *Server key* is created after removing clues from each corner cell. *User key* is created by storing all the removed corner digits (row-wise); if there is no clue, then 0 is stored as the corresponding key of the cell. For example, in Fig. 7, the user key we obtain is 0850, as there are no clues in the top-left corner cell as well as bottom-right corner cell; we may note that the top-right corner cell contains an 8 and the bottom-left corner cell contains a 5 as clues. In our scheme, we like to keep the server key is retained in the database server, while the user key is returned back to the user for future authentication, as and when necessary.

Then, the biometric sketch is created, and this sketch along with the server key is stored in the server. So, in our proposed scheme, the server stores encrypted biometric template sketches in the database as well as the server key generated using the sudoku instance. The user key is kept by the biometric user. If anybody wants to enter into the biometric system, one has to supply the biometric information as well as the key generated by the server. The decryption technique of the proposed scheme is as follows.

4.1 Biometric Template Decryption Technique

Step 1   *Submission of keys*: User has to submit its own key along with biometric data to the server. The server places each number sequentially to each corner of the server key as shown in Fig. 8. The first leftmost value is placed in the top-left corner of the server key. The next value is placed in the top-right corner, whereas the next two values are placed in the bottom-left and bottom-right corner of the server key, respectively. Then, the original instance of the sudoku is computed, wherefrom we can reach to the original solved sudoku puzzle, which is ultimately embedded in the template. If 0 is found in the user key, then the value is replaced by a blank cell in the server key.

Fig. 8 Merging of Server key with user key

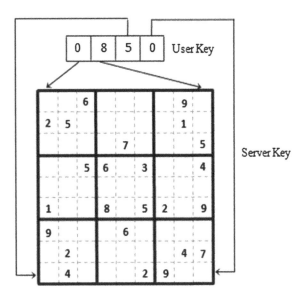

Step 2 Here, the sudoku instance is solved by the server to get the complete sudoku puzzle.

Step 3 Then, in the similar way during the encryption process, the reconstructed templates are divided into 81 blocks.

For each individual block:
Make groups of four pixels each.
For each group:
The associated value present in a sudoku cell is subtracted from the LSB of each pixel present in the corresponding group of the block, and then, padding is also removed, if added earlier.

Then, the original biometric template (i.e., X, in Fig. 1) is recreated.

Now, this is obvious that an efficient sudoku solving algorithm is also very essential to speed up the process of encryption and decryption. So for this purpose, in our proposed work we have also designed an efficient sudoku solving algorithm, which is described in the next section.

4.2 The Designed Sudoku Solving Algorithm

Our devised algorithm considers each of the minigrids that may be numbered as 1 through 9 as shown in Fig. 9. Each minigrid may or may not have some clues as numbers that are given. We first consider a minigrid that contains a maximum number of clues, and if there are two or more such minigrids, we consider the one with the least minigrid number.

Needless to mention that each of the cells in a minigrid, either containing a clue or a blank cell, is somehow differentiated from each of the cells of another minigrid as the position of a cell in a sudoku instance could be specified by its row number and column number, which is unique. So, a cell $[i, j]$ of minigrid k may either contain a number l as a given clue or a blank location that is to be filled in by inserting a number m, where $1 \leq i, j, k, l, m \leq 9$.

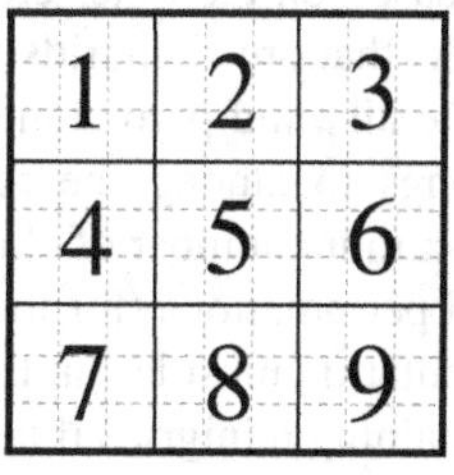

Fig. 9 The structure of a 9×9 sudoku puzzle (problem) with its nine minigrids of size 3×3 each as numbered 1 through 9. Minigrid number 1 consists of the cell locations [1], [1, 2], [1, 3], [1, 2], [2], [2, 3], [1, 3], [2, 3], and [3], minigrid number 2 consists of the cell locations [1, 4], [1, 5], [1, 6], [2, 4], [2, 5], [2, 6], [3, 4], [3, 5], [3, 6], and so on

Now to start with a minigrid as stated above, we find that the minigrid 3 contains a maximum number of clues, i.e., four, among all the minigrids, and each of the minigrids 1 and 2 contains less number of clues than that of minigrid 3 (see Fig. 2a). For example, for the sudoku instance as shown in Fig. 2a, each of the minigrids 3, 5, and 7 contains four clues each; hence, at the beginning, we consider minigrid 3 for computing all its valid permutations of the missing numbers for its blank locations (as 3 is the minimum minigrid number).

Besides, for a given sudoku instance, we know all the clues given as well as the clue positions among the cells of a minigrid and subsequently the blank cells are also known to us. For example, the given clues in minigrid 3 of Fig. 2a are 9 at location [1, 8], 8 at location [1, 9], 1 at location [2, 8], and 5 at location [3, 9]. Here, we denote a cell location of a sudoku instance by [row number, column number], where each of row numbers and column numbers varies from 1 to 9. Hence, the blank locations are [1, 7], [2, 7], [2, 9], [3, 7], and [3, 8], and the missing digits are 2, 3, 4, 6, and 7.

We compute all possible permutations of these missing digits in minigrid 3, where the first permutation may be 23467 (the minimum number) and the last permutation may be 76432 (the maximum number using the missing digits). Here, as the number of blank locations is five, the total number of permutations is 5!, which is equal to 120. Now, the algorithm considers each of these permutations one after another and identifies only the valid set of permutations based on the given clues available in rows and columns in other minigrids (that are minigrids 1, 2, 6, and 9). As for example, if we consider the first permutation 23467 and place the missing digits, respectively, in order in locations [1, 7], [2, 7], [2, 9], [3, 7], and [3, 8], which are arranged in rising mode, we find that this permutation is not a valid permutation. This is because the location [6, 7] already contains 2 as a clue of minigrid 6, and we cannot place 2 at [1, 7] as the permutation suggests. Also the location [3, 5] contains 7 as a clue of minigrid 2, and we cannot place 7 at [3, 8] as it is supposed to place.

Similarly, we may find that the last permutation 76432 is also not a valid permutation as location [4, 9] already contains 4 as a clue of minigrid 6, and we cannot place 4 at [2, 9] as the permutation suggests. But we may observe that 74362 is a valid permutation as we may safely place 7 at [1, 7], 4 at [2, 7], 3 at [2, 9], 6 at [3, 7], and 2 at [3, 8] based on the other clues in the corresponding rows and columns of other minigrids (that are minigrids 1, 2, 6, and 9).

This is how we may compute all valid permutations of minigrid 3 and proceed for a next minigrid that belongs to among the row and column minigrids of minigrid 3 which contains a maximum number of clues, but the minigrid number is minimum. Among all the valid permutations (for their respective blank locations) of minigrid 3, at least one permutation must last at the end of computation of valid permutations of each of the remaining minigrids if the solution of the given sudoku instance is unique. To find out the next minigrid to be considered, we go through the row and column minigrids of minigrid 3 in the sudoku instance of Fig. 2a (that are minigrids 1, 2, 6, and 9), and among these minigrids, we find that the minigrid

1 contains a maximum number of clues, i.e., three (which is equally true for each of the minigrids 6 and 9), and its minigrid number is the minimum.

So now, we consider minigrid 1, and as done before for minigrid 3, we find the given clues and the missing digits therein along with their locations. Here, we do exactly the same as we did earlier in computing all permutations of the missing digits in minigrid 3. At the time of identifying all valid permutations of minigrid 1, we consider one valid permutation (at their respective blank locations) of minigrid 3 in addition to all given clues of the instance under consideration. If we get at least one valid permutation for minigrid 1 (obeying an assumed valid permutation of minigrid 3), we consider it for some subsequent computation of permutations of another minigrid; otherwise, we consider a second valid permutation of minigrid 3, and based on that, we compute another set of valid permutations for minigrid 1, and so on.

Now, it is straightforward to declare that here, the minigrid that is to be considered is one among the minigrids 2, 4, 6, 7, and 9 as the row and column minigrids of minigrids 3 and 1 (for which we have already computed valid permutation(s) one after another); note that neither of minigrids 5 and 8 are a row or column minigrid of minigrids 3 and 1. Hence, following the instance in Fig. 2a, we consider minigrid 7 for computing all its valid permutations allowing for one valid permutation of minigrid 3 and then one subsequent valid permutation of minigrid 1, in addition to all given clues of the instance under consideration, as each of the minigrids 2, 4, 6, and 9 contains less number of clues than that of minigrid 7. Here in computing all valid permutations of minigrid 7, we may not consider an assumed valid permutation of minigrid 3, as this minigrid is neither in a row nor in a column of minigrid 7, but we have to consider a valid permutation of minigrid 1 and all given clues in the sudoku instance (primarily the clues given in minigrids 4, 8, and 9).

This process is continued till a valid permutation of a minigrid (or a set of valid permutations of a group of minigrids) is propagated to compute a valid permutation of a subsequent minigrid, and eventually, a valid permutation of the last minigrid (i.e., the ninth minigrid; not necessarily minigrid number 9) is computed, which altogether generate a desired solution of the given sudoku instance. It may so happen that up to t minigrids, t valid permutations that we consider in a series match each other toward a valid combination of the given sudoku instance, but there is no valid permutation for the $(t + 1)$th minigrid obeying the earlier assumed t valid permutations, where $1 < t < 9$. Then, we consider a second valid permutation of the tth minigrid, and after that we try to compute a valid permutation for the $(t + 1)$th minigrid, if one exists. If for none of the valid permutations of the tth minigrid a valid permutation for the $(t + 1)$th minigrid is obtained, we consider a second valid permutation for the $(t - 1)$th minigrid that leads to compute a new set of valid permutations for the tth minigrid, and so on.

We claim that we must acquire at least one valid permutation for each of the minigrids one after another, obeying at least one valid permutation computed for each of the minigrids considered earlier in the process of assuming the minigrids in succession; we claim this result in the form of the following theorem if at least one solution of the given sudoku puzzle exists.

Theorem 1 *There is at least one valid permutation for the missing digits for their respective blank locations in each of the minigrids such that the combination of all such (nine) valid permutations for all the (nine) minigrids produces a desired solution, if there exists a solution of a given sudoku instance.*

Proof The verification of the theorem is straightforward following the steps of the inherent development of the algorithm as stated above, if a feasible solution of the given sudoku instance is there. We may start with one valid permutation for some earlier assumed minigrid that may not be a valid partial solution in combination for the whole sudoku instance; then, we must reach to a point of computing a valid permutation of some subsequent minigrid when no such permutation is obtained for that minigrid. In that case, we are supposed to return back to the former minigrid we had to consider a next valid permutation, if any, for the same (i.e., for the previous minigrid) and move to the current minigrid for computing its valid permutations accordingly. Hence, it is clear that if one valid permutation for some earlier assumed minigrid is not a valid partial solution in combination for the whole sudoku instance, then we must have to return back to that prior minigrid to consider a new valid permutation of the same to continue the process again in computing all valid permutations of its subsequent minigrid, and so on. In this way, a set of individual valid permutations is to be differentiated so that in combination of all of them a desired solution of the given sudoku instance is computed, if one such solution exists.   □

To see the algorithm at a glance, let us write it in the form as follows:
Input: A sudoku instance, P of size 9×9.
Output: A solution, S of the given sudoku instance, P.

Step 1   Compute the number of clues, digit(s) given as clue, and the missing digits in each of the minigrids of P.

Step 2   Compute S_M, a sequence of minigrids that contains all the minigrids in succession, wherein $M \in S_M$ is the minigrid (and the first member in S_M) with a maximum number of clues and whose minigrid number is minimum. In S_M, a member N is a minigrid which is either in the row or in the column of any of its earlier members in S_M including M that contains a maximum number of clues and whose minigrid number is minimum, where $1 < N \leq 9$.

Step 3   Compute all valid permutations for the missing digits in M and store them.

Step 4 For all the remaining minigrids in succession in S_M do the following:

Step 4.1: Consider a next minigrid, $N \in S_M$, and compute all its valid permutations for the missing digits in N assuming a valid permutation for each of the earlier minigrids up to M, and store them.

Step 4.2: If one valid permutation for N is obtained, then consider a next minigrid of N in S_M, if any, and compute all its valid permutations for the missing digits in this minigrid assuming a valid permutation for each of the earlier minigrids up to M, and store them.

Else consider a next valid permutation, if any, of the immediately previous minigrid of N, and compute all its valid permutations for the missing digits in N assuming a valid permutation for each of the earlier minigrids up to M, and store them.

Step 5 If all the valid permutations of the immediate successor minigrid of M are exhausted to obtain a valid combination for all the nine minigrids in S_M, then consider a next valid permutation of M and go to Step 4. The process is continued until a valid combination for all the nine minigrids in S_M is obtained as a desired solution S for P; otherwise, the algorithm declares that there is no valid solution for the given instance P.

Now it is straightforward to compute S_M for a given sudoku instance P. As for example, consider the sudoku instance given in Fig. 2a. According to this instance, the sequence S_M of minigrids is $\langle 3, 1, 7, 6, 5, 9, 4, 8, 2 \rangle$ as it has been described and performed in Step 2 of the first version of the algorithm above.

Computation of all valid permutations for the missing digits in a minigrid is an important task of the present algorithm. At the time of computing only all valid permutations for the missing digits, we follow a tree data structure, where the degree of the root of the tree is same as the number of missing digits, and level-wise it reduces to one to obtain the leaf vertices, where each leaf at the lowest level is a valid permutation of all the missing digits based on the clues given in P (and the assumed valid permutation(s) in other minigrid(s) in subsequent iterations).

As for example, the number of clues given in minigrid 3 of the puzzle in Fig. 2a is four, and the missing digits are 2, 3, 4, 6, and 7. The proposed algorithm likes to place each of the permutations of these missing digits in the blank locations [1, 7], [2, 7], [2, 9], [3, 7], and [3, 8]. Here, the tree structure we like to compute is shown in Fig. 10, whose root does not contain any permutation of the missing five digits, and it is represented by '*****'. This root is having five children where the first child leads to generate all valid permutations staring with 2, the second child leads to generate all valid permutations staring with 3, and so on.

Now note that none of the permutations starting with 2 is a valid permutation as column 7 of minigrid 6 contains 2 as given clue (at location [6, 7]). So, we do not expand this vertex (i.e., vertex with permutation '2***') further in order to

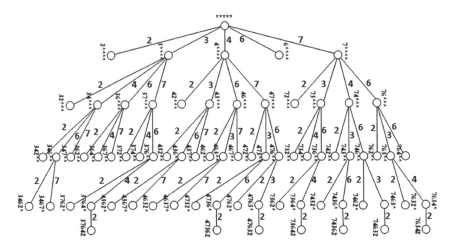

Fig. 10 The permutation tree for generating only valid permutations of the missing digits in minigrid 3 of the sudoku instance shown in Fig. 2a

compute only the set of desired valid permutations. Similarly, we do not expand the child vertex with permutation '6****', as location [1, 3] contains 6 as given clue. Up to this point in time, as either 3, or 4, or 7 could be placed at [1, 7], we expand each of the child vertices starting with permutations 3, and 4, and 7, as shown in Fig. 10.

Similarly, we expand the tree structure inserting a new missing number at its respective location (for a blank cell) leading from a valid permutation (as vertex) in the previous level of the tree. Correspondingly, we verify whether the missing digit could be placed at the respective location for a blank cell of the given sudoku instance P. If the answer is 'yes,' we further expand the vertex; otherwise, we stop expanding the vertex in some earlier level of the tree structure prior to the last level of leaf vertices only. As for example, the vertex with permutation '742**' is not expandable, because we cannot place 2 at [2, 9] as [1, 2] contains a 2 as given clue. So, this is how either a valid permutation is generated from the root of the tree structure reaching to a bottommost leaf vertex, or the process of expansion is terminated in some earlier level of the tree that must generate other than valid (unwanted) permutations at this point in time.

Interestingly, Fig. 10 shows the reality that the number of possible permutations of five missing digits is 120, and out of them, only seven are valid for minigrid 3 of the sudoku instance shown in Fig. 2a. Note that the given clues in P are nothing but constraints, and we are supposed to comply with each of them. So, usually, if there are more clues, P is more constrained, and hence, the number of valid permutations is even much less, and the solution, if it exists, is unique in most of the cases. On the contrary, if there are fewer clues in P, more valid permutations for some minigrid of P could be generated, computation of a solution for P might take more time, and P may have two or more valid solutions. In any case, if there is a unique solution of the assumed sudoku instance (in Fig. 2a), out of these seven valid permutations, only one is finally be accepted following the subsequent steps of the algorithm.

Now, the algorithm considers one valid permutation (out of the seven permutations) of minigrid 3 and all given clues in P and generates all valid permutations for minigrid 1. If at least one valid permutation for minigrid 1 is obtained, we proceed for generating all valid permutations for minigrid 7 obeying all given clues in P and the assumed valid permutations of minigrids 3 and 1; otherwise, a second valid permutation of minigrid 3 is considered, for which in a similar way, we generate all valid permutations for minigrid 1, and so on.

This is how the algorithm proceeds and generates all valid permutations of a minigrid under consideration conforming the given clues in P and a set of assumed valid permutations, one for each of the minigrids considered earlier in succession, up to this point in time.

Note that at the time of computing a set of valid permutations for a minigrid, we have to consider clues and (earlier computed) valid permutations in only four of the remaining eight minigrids that are adjacent to the minigrid (currently) under consideration. As for example, while computing valid permutations for minigrid 7, we have to consider one valid permutation of minigrid 1 and the clues given in minigrids 1, 4, 8, and 9 only; here, the assumed valid permutation of minigrid 3 has no use while computing valid permutations for minigrid 7. In the same way, while computing valid permutations for minigrid 6, only we have to consider the assumed valid permutation of minigrid 3 (up to this point in time) and the clues given in minigrids 3, 4, 5, and 9 only; here, the assumed valid permutations of minigrids 1 and 7 have no use while computing valid permutations for minigrid 6 and so on.

Now, we discuss about the size of the tree structure under consideration. If p be the number of blank cells in a minigrid and the sudoku instance is of size $n \times n$, then the computational time as well as the computational space complexity of the sudoku solver developed herein is $(p! - x)^n = O(p^n)$, where x is the number of other than valid permutations based on the clues given in the sudoku instance P. Our observation is that for a given sudoku instance P, x is very close to $p!$, and hence, $p!-x$ is a reasonably small number and in our case the value of n is equal to 9. Hence, the experimentations made by this algorithm take negligible amount of clock time, and this is of the order of milliseconds.

5 Experimental Results: Analysis and Discussion

We have tested our proposed approach of biometric template encryption scheme, using the Essex Faces 94 face database (E94 database), which is publicly available and essentially created for face recognition related research studies. The E94 database contains images of 152 distinct subjects, with 20 different images for each subject where the size of each JPEG image is 180×200. In our approach, we have transformed these images to 8-bit gray-level images and then used these gray-

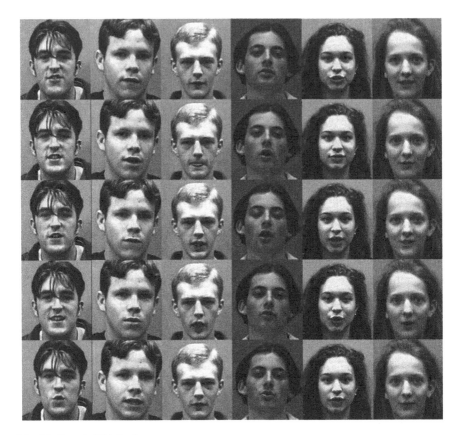

Fig. 11 Sample E94 database

level images in our experiments. For each subject, we have randomly chosen the 12 out of 20 samples for enrollment and the remaining 8 sample face images are used for authentication. Some sample images from the E94 database are given in Fig. 11.

Then, we have done histogram analysis of this proposed scheme. We have noticed a very less distortion in the template. Sample histograms are shown in Figs. 12 and 13; Fig. 12 shows the histogram of the template before embedding sudoku, and Fig. 13 shows the same after embedding sudoku.

Fig. 12 Histogram of the template before embedding sudoku

Fig. 13 Histogram of the template after embedding sudoku

From the histogram analysis, we can easily notice that there is incredibly less distortion in the image after embedding the sudoku puzzle, which is one of the prime requirements of any template encryption scheme.

5.1 Computation Time

As we have performed only simple add operation on the LSB's, it takes negligible amount of computational time. The average computational time for the operation is around 30 ms, which is small enough as compared to other encryption techniques. Feature transformation takes in an average of 30–35 ms, whereas biometric cryptosystem takes an average of 45–45 ms of computation time.

5.2 Robustness

In feature transformation approach, a noninvertible feature transformation function is applied to the biometric template, whereas in case of biometric cryptosystems using keys, biometric templates are embedded. But still there are scopes to modify the template which ultimately leads to denial-of-service (DOS) attack. As in our proposed scheme, we are embedding a sudoku puzzle inside the biometric template and each 9×9 sudoku puzzle having the number 1–9 in each row, column, and minigrid only once. Modification is not possible, as any modification in the template ultimately leads to modification in the sudoku puzzle itself, which leads to the violation of sudoku constraints.

5.3 More Number of Keys

There exist as many as 6,670,903,752,021,072,936,960 distinct sudoku puzzles [2]. That means these many numbers of different keys can be used. Now, the average computing time of a sudoku instance is $\sim$29 ms; that means, around 6,134,456,139,289 years are required to solve all the sudoku instances available in practice. So guessing the key is almost impossible in our proposed method, and the brute force attack is also not feasible in our projected scheme of encryption.

6 Conclusion

Our anticipated encryption scheme is novel in the following sense. As we are embedding a sudoku puzzle inside the template, it is almost impossible for an intruder to modify the template as each row, column, and minigrid of the sudoku instances contains 1–9 uniquely. Any changes in these data ultimately lead to an error in the sudoku. If somebody changes the whole sudoku puzzle, then also the user of the system is capable to find the alteration, as the original sudoku instance is stored as key in the server as well as with the user. We are modifying the LSB of each pixel. Hence, the probability of image distortion is significantly reduced. We have distributed the key used for the encryption into two parties, i.e., user and server. So, without getting these two keys, nobody is able to decrypt the template. Therefore, we can claim that the proposed scheme is sincerely robust with least distortion in the quality of the template.

References

1. www.ftc.gov/opa/reporter/idtheft/index.shtml
2. Roberts, C.: Biometric attack vectors and defenses. Comput. Secur. **26**(1), 14–25 (2007)
3. Jain, A.K., Ross, A., Uludag, U.: Biometric template security: challenges and solutions. In: Proceedings of the European Signal Processing Conference (EUSIPCO '05), Antalya, Turkey, Sept 2005
4. Cukic, B., Bartlow, N.: Biometric system threats and countermeasures: a risk based approach. In: Proceedings of the Biometric Consortium Conference (BCC '05), Crystal City, USA, Sept 2005
5. Jain, A.K., Ross, A., Pankanti, S.: Biometrics: a tool for information security. IEEE Trans. Inf. Forensics Secur. **1**(2), 125–143 (2006)
6. http://www.en.wikipedia.org/wiki/Pentomino
7. http://www.en.wikipedia.org/wiki/Tetris
8. Jussien, N.: A–Z of Sudoku. ISTE Ltd., USA (2007)
9. Lee, W.-M.: Programming Sudoku. Apress, USA (2006)
10. Maltoni, D., Maio, D., Jain, A.K., Prabhakar, S.: Handbook of Fingerprint Recognition. Springer, Berlin (2003)
11. Uludag, U., Pankanti, S., Prabhakar, S., Jain, A.K.: Biometric cryptosystems: issues and challenges. Proc. IEEE **92**(6), 948–960 (2004)
12. Cavoukian, A., Stoianov, A.: Biometric encryption: a positive-sum technology that achieves strong authentication, security, and privacy. Technical Report, Office of the Information and Privacy Commissioner of Ontario, Toronto, Ontario, Canada, March 2007
13. Vetro, A., Memon, N.: Biometric system security. In: Proceedings of the Second International Conference on Biometrics, Seoul, South Korea, Aug 2007

Part III
Computer Aided Design

An ESOP-Based Reversible Circuit Synthesis Flow Using Simulated Annealing

Kamalika Datta, Alhaad Gokhale, Indranil Sengupta
and Hafizur Rahaman

Abstract The problem of reversible circuit synthesis has become very important with increasing emphasis on low-power design and quantum computation. Many synthesis approaches for reversible circuits have been reported over the last decade. Among these approaches, those based on the exclusive-or sum-of-products (ESOP) realization of functions have been explored by many researchers because of two important reasons: large circuits can be handled, and the mapping from ESOP cubes to reversible gate netlist is fairly straightforward. This paper proposes a simulated annealing (SA)-based approach for transforming the ESOP cubes generated from Exorcism-4 tool using some cube mapping rules, followed by a strategy to map the ESOP cubes to a netlist of reversible gates. Both positive- and negative-control Toffoli gates are used for synthesis. Synthesis results on a number of reversible logic benchmarks show that for many of the cases, it is possible to get a reduction in quantum cost against the best-known methods.

Keywords Reversible circuits · ESOP · Simulated annealing · Template matching

K. Datta (✉) · H. Rahaman
Department of Information Technology, Indian Institute of Engineering Science
and Technology, Shibpur, Howrah 711103, India
e-mail: kdatta.iitkgp@gmail.com

H. Rahaman
e-mail: hafizur@vlsi.becs.ac.in

A. Gokhale · I. Sengupta
Department of Computer Science and Engineering, Indian Institute of Technology,
Kharagpur 721301, India
e-mail: alhaadgokhale@gmail.com

I. Sengupta
e-mail: isg@iitkgp.ac.in

© Springer India 2015 131
R. Chaki et al. (eds.), *Applied Computation and Security Systems*, Advances in Intelligent
Systems and Computing 305, DOI 10.1007/978-81-322-1988-0_8

1 Introduction

With great advancements in semiconductor technology over the last few decades, the number of transistors in a chip has grown exponentially, and the age-old Moore's law [14] continues to hold. With such miniaturization, power dissipation has become a major problem with today's VLSI chips. Various low-power design techniques and architectures have been proposed to counter these problems.

Landauer [11] showed that whenever there is loss of information during some computation, energy is dissipated in the form of heat. This has been quantified as KT log 2 J of energy for every bit of information that is lost, where K is the Boltzmann constant and T is the absolute temperature of the environment. Landauer's principle has also been experimentally verified [3], by actually measuring the energy dissipated when one bit of information is erased. Since traditional irreversible gates lose information during computation, they will always dissipate energy irrespective of the underlying technology. However, since reversible computations are information lossless, they have the potential for having very low-power implementations. This conjecture is also supported by an observation by Bennett [2], who argued that zero power dissipation is possible only if the computation is information lossless (that is, reversible). Moreover, reversible computing finds its importance in quantum computation as well, where the basic operations are reversible in nature.

Recently in [23], the authors proposed a reversible implementation of a low-power channel encoding scheme and showed that the corresponding CMOS realization consumes less power as compared to the best-known conventional encoding method. Again in [5], a reversible implementation of the AES encryption algorithm has been proposed. This ongoing effort by various researchers aims to establish reversible computing as an alternate low-power design paradigm.

With such motivations, synthesis of reversible circuits has become an active area of research. Various synthesis approaches have been proposed, which can be broadly classified as exact methods [9], heuristic methods [4], and those based on higher-level function representations [7, 22]. Exact (heuristic) methods generate optimal (near-optimal) circuits but cannot handle large functions. Methods based on higher-level function representations such as binary decision diagrams (BDD) and exclusive-or sum-of-products (ESOP), in contrast, are able to synthesize large circuits with several hundreds of inputs, again with no guarantee of optimality. ESOP-based synthesis methods have the added advantage that they can also handle nonreversible or incompletely specified functions, with the input given as a .pla file.

The cost metrics that are typically used to evaluate a synthesized gate netlist are number of gates, quantum cost [1], or the number of equivalent transistors. Here, we have used the quantum cost metric for all comparison and evaluations.

In this paper, an integrated approach to synthesis of reversible circuits is presented, which is based on the ESOP representation of functions. A simulated annealing (SA)-based approach is proposed for transformation of ESOP cubes,

which are then mapped to a reversible gate cascade consisting of both positive-
and negative-control Toffoli gates. The rest of the paper is organized as follows.
Section 2 presents a review of the ESOP-based synthesis methods and some of the
important issues therein. Section 3 gives the theoretical framework based on which
the cube transformations are carried out. Section 4 explains the proposed scheme.
The experimental results are summarized in Sect. 5, with Sect. 6 giving some
concluding remarks and some scopes for future work.

2 Background of ESOP-Based Synthesis

In this section, we will briefly review the basics of reversible logic and ESOP-
based synthesis.

2.1 Reversible Logic and Reversible Gates

A Boolean function $f : \mathbf{B}^n \to \mathbf{B}^n$ is said to be reversible if there is a one to one
mapping and it is bijective. The problem of synthesis is to determine a reversible
circuit that realizes a given function f.

Like in many previous methods, in this paper, we consider the gate library
consisting of NOT, CNOT, and generalized Toffoli gates. The method uses generalized Toffoli gates with both positive and negative controls. Figure 1 shows
CNOT, positive-control, and negative-control Toffoli gates that realize the func-
tions: $\{a, a \oplus b\}$, $\{a, b, c \oplus ab\}$, and $\{a, b, c \oplus \bar{a}\bar{b}\}$, respectively.

To estimate the cost of an implementation, several metrics are used, namely
number of gates, number of equivalent MOS transistors, and number of equivalent
basic quantum operations called quantum cost [1]. There are standard ways of
computing the quantum cost from a given gate netlist [6], for gates having positive
controls only. Recently in [12], Toffoli gates with negative controls have been
introduced. For calculating quantum cost, the same calculation for positive controls will hold for negative controls as well, with the only exception for the case
where all the controls are negative. In such case, a 1 has to be added to the
quantum cost as calculated.

2.2 ESOP-Based Synthesis Techniques

Exclusive-or sum-of-products (ESOP) is a type of representation of a Boolean
function, as an exclusive-or sum of several product terms (called cubes). An
example function in ESOP form is $f = ab \oplus cd \oplus \bar{a}\,\bar{c}d$. One interesting thing

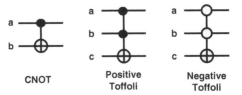

Fig. 1 Basic reversible gates

about the ESOP representation is that we can map the ESOP cubes into equivalent Toffoli gates in a straightforward way [7]. The ESOP cube list for a given function can be generated by one of several ESOP generation algorithms, but in most of the papers, a tool called Exorcism-4 [13] is used.

In [10], Gupta et al. proposed a synthesis approach based on positive-polarity Reed–Muller expansion, which is a form of ESOP. Here instead of using a single gate for each term in the expansion, a tree-like structure is used to explore all possible factors of each term, and in turn, a circuit is constructed that shares factors.

In [7], Fazal et al. presented a synthesis approach, where a Toffoli gate is directly generated from an ESOP cube list. The total number of lines required is $(2n + m)$ where n and m denote the number of input and output variables, respectively. A Toffoli gate is added for each output of a particular cube. In a further modification, the number of lines was reduced to $(n + m)$ by inserting a few NOT gates.

In [19], Sanaee et al. proposed a method that exploits the sharing of cubes among outputs. A cube is realized once, and the result is transferred to the other sharing outputs using CNOT gates. In [20], the authors proposed another technique that uses negative-control Toffoli gates and some transformation rules to reduce some of the output lines.

In [18] Rice et al. proposed an approach which uses an autocorrelation-based cost metric for identifying the position of Toffoli gates. This method requires more number of gates for many circuits as compared to [7].

In [15], Nayeem et al. presented a shared cube-based approach, which tries to optimize the Toffoli gate mapping by grouping the cubes into sublists. They achieved a significant reduction in quantum cost as compared to other works.

In [17] Rice et al. proposed an ordering-based technique to reorder the ESOP cube list to reduce the number of NOT gates. The method shows better results as compared to [7, 18].

In [6, 8], the authors suggest approaches to optimize the ESOP cubes using the pseudo-Kronecker representation of a Boolean function. An evolutionary algorithm is proposed in [6] to determine a good variable ordering and a suitable XOR decomposition for the BDD, so as to minimize the quantum cost.

This paper proposes a rule-based technique for transforming a set of ESOP cubes using simulated annealing, with the objective of reducing the quantum cost of the corresponding reversible gate implementation. The theoretical basis of the work is discussed in the following section.

3 Theoretical Framework for the Work

In this section, we discuss some methods for transforming the ESOP cubes such that the quantum cost of the resulting netlist is reduced. Some of these rules were used in earlier works in the context of synthesis; however, in the present work, we use them for cube transformation. The following results can be easily proved using switching algebra.

Lemma 1 *For an n-variable function, if A, B and Z denote cubes such that Z does not contain any variables present in A or B, then*

$$(A \oplus B)Z = AZ \oplus BZ$$

Proof We have

$$AZ \oplus BZ = AZ(\overline{B} + \overline{Z}) + BZ(\overline{A} + \overline{Z})$$
$$= A\overline{B}Z + \overline{A}BZ$$
$$= (A\overline{B} + \overline{A}B)Z = (A \oplus B)Z$$

Theorem 1 *For an n-variable function, if A, B, C, D and Z represent cubes such that Z does not contain any of the variables present in A, B, C or D, then the following result holds:*

$$\text{if } A \oplus B = C \oplus D, \text{ then } AZ \oplus BZ = CZ \oplus DZ$$

Proof Let us assume that $A \oplus B = C \oplus D$. Therefore,

$$AZ \oplus BZ = AZ(\overline{B} + \overline{Z}) + BZ(\overline{A} + \overline{Z})$$
$$= A\overline{B}Z + \overline{A}BZ = (A\overline{B} + \overline{A}B)Z$$
$$= (A \oplus B)Z = (C \oplus D)Z$$
$$= C\overline{D}Z + \overline{C}DZ$$
$$= CZ(\overline{D} + \overline{Z}) + DZ(\overline{C} + \overline{Z})$$
$$= CZ \oplus DZ$$

Theorem 2 *For an n-variable function, if A_i $(1 \le i \le p)$, B_i $(1 \le i \le m)$, and Z represent cubes such that Z does not contain any of the variables present in A_i or B_i, then the following result holds:*

$$\text{If } A_1 \oplus A_2 \oplus \ldots \oplus A_p = B_1 \oplus B_2 \oplus \ldots \oplus B_m,$$
$$\text{then } A_1Z \oplus A_2Z \oplus \ldots \oplus A_pZ = B_1Z \oplus B_2Z \oplus \ldots \oplus B_mZ.$$

Proof Follows along the same lines as in the previous theorem.

3.1 Cube Transformation Rules

A set of rules that can be used to transform a set of ESOP cubes is presented below. The basic idea is to transform a set of ESOP cubes into an equivalent set of cubes through selective application of these rules so that the quantum cost of the final gate netlist is reduced.

R1: Rice and Nayeem [17] if two cubes A and B differ in one position, where it is '1' in A and '–' in B, then we merge them into a single cube C by setting the differing position to '0'. For example, the cubes {110–, 1–0–} can be merged to {100–}.

R2: Rice and Nayeem [17] if two cubes A and B differ in one position, where it is '0' in A and '–' in B, then they can be merged by setting the differing position to '1'. For example, the cubes {10–, 1–} can be merged to {11–}.

R3: Mishchenko and Perkowski [13] if two cubes A and B differ in one position, where it is '0' in A and '1' in B, then we merge them by setting the differing position to '–'. For example, the cubes {10–, 11–} can be merged to {1–}.

R4: Mishchenko and Perkowski [13] split a cube into two cubes on a '–' and make it '0' in one of the cubes, and '1' in the other. For example, the cube {1–0–} can be split into the pair of cubes {100–, 110–}.

R5: Using EXOR-link operation [13], a pair of cubes that are distance k apart can be replaced by a set of k cubes of larger sizes. For example, the cubes {000, 111} that are distance-3 apart can be replaced by the cubes {00–, –01, 1–1}.

R6: If there are two cubes A and B that are at a distance of 2 apart, add two copies of a cube C which is at a distance of 1 from both A and B and then merge (A, C) and (B, C) using rules R1, R2 or R3.
For example, the cubes {1010, 0011} are at a distance of 2 apart. We select a cube {1011} that is at unit distance from both. Then, we merge (1010, 1011) and (0011, 1011) to get the cubes {101–, –011}.

R7: If there is a set of cubes at even distances from each other, then repeated applications of rule R6 followed by rules R1, R2 or R3 can be used to achieve reduction in quantum cost.
For example, consider the cubes {1010, 0011, 1100, 1111} every pair of which are either at distance 2 or 4 from each other. Using rule R7, we merge cubes as: (1010, 0011) = (101–, –011) and (1100, 1111) = (11–0, 111–).

Using rule R3, we can merge (101–, 111–) = (1–1–). We thus get the final set of cubes as: {–011, 11–0, 1–1–}.

R8: Split two cubes at a distance of 3 into a set of three larger cubes. For example, {101, 010} = {1–1, 11–, –10}.

The results of Theorems 1 and 2 can be utilized to extend the applicability of the above rules, as shown in the example below.

Example 1 The cubes {101–10, 010–10} can be transformed into {1–1–10, 11–10, –10–10} by using rule R10. Here, Z is '–10'. Similarly, the cubes {101–, 011–} can be transformed into {1–1–, –11–} using rule R9. Here, Z is '1–'.

□

4 The Proposed Synthesis Approach

The proposed cube transformation and gate mapping approach is discussed in this section. The steps used in the synthesis process are listed below:

1. Firstly, the input function specification is provided as a .pla file, which is transformed into a set of ESOP cubes (in .esop format) using the Exorcism-4 tool [13].
2. Then, the proposed simulated annealing-based cube transformation tool is used to modify the ESOP cubes into more desirable forms (with respect to quantum cost) and generate another .esop file.
3. The final .esop cube list is mapped to reversible gate cascade consisting of positive- and negative-control generalized Toffoli gates using optimization concepts as proposed in [7, 17, 20], along with some heuristics suggested in Sect. 4.2.

A block diagram of the synthesis flow is depicted in Fig. 2, which also shows an optional last step of template-based optimization with both positive- and negative-control gates to reduce the number of gates and also the quantum cost.

4.1 Cube Transformation Using Simulated Annealing

In this step, the cube list generated using Exorcism-4 is transformed into a more desirable form by selective application of the rules presented in the previous section. First, an approach where the rules are iteratively applied to a cube list as

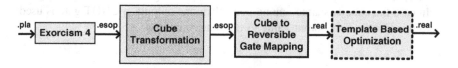

Fig. 2 The overall synthesis approach

long as there is a reduction in quantum cost was tried out. However, results were not so encouraging as the algorithm was often getting stuck in a local minimum. For this reason, a simulated annealing-based approach has been tried out that allows worse moves during the initial phases of the iterative process, while exhibiting greedy behavior toward the later stages. The pseudo-code of the algorithm is stated below.

Algorithm Simulated annealing-based cube reordering

> **Input:**   An initial ESOP cube list I
> **Output:** The transformed ESOP cube list F
> **begin**
>> $F = I$;
>> $C = compute\_cost\ (F)$;
>> $T = 100000$;   // Initial temperature
>> **do**
>>> **for** $i = 1$ **to** 1000 **do**
>>> **begin**
>>>> $r = choose\_rule\ (F)$;   // choose a rule randomly
>>>> $F_{new} = trial\_move\ (F,\ r)$;
>>>> $C_{new} = compute\_cost\ (F_{new})$;
>>>> **if** $(C_{new} < C)$
>>>>> $F = F_{new}$;
>>>> **else**
>>>>> **if** $(e^{(C-C_{new})/T} > \mathrm{random}())$
>>>>>> $F = F_{new}$;
>>> **end**
>>> $T = T * 0.1$;
>> **while** $(T > 0.01)$;
> **end**

The algorithm may accept worse moves at higher values of T, but becomes more greedy as the iteration proceeds and T becomes less. The parameters have been tuned through extensive experimentation. A fast cooling scheme is chosen (i.e., $T = T \times 0.1$), since it is observed that slower cooling sequences do not give any better results. The number of iterations in every cooling cycle is also selected in a similar way.

- The function *choose_rule* (F) scans the cube list F, randomly selects a rule r (vide Sect. 3) that can be applied to F, and returns the rule number r.
- The function *trial_move* (F, r) applies the rule r to the cube list F and returns the modified cube list.
- The function *compute_cost* (E) estimates the quantum cost of a given cube list E, by transforming every cube into a positive- or negative-control Toffoli gate. In case the cube is shared among more than one outputs, a CNOT gate is used to transfer the first output value to the other output(s).

4.2 Cube to Reversible Gate Mapping

There exists various techniques in the literature for mapping ESOP cubes into reversible gate netlists [7, 15, 20]. In the present work, we have used a combination of several techniques along with some heuristic gate mapping rules, with the objective of obtaining a lower quantum cost. The ESOP cubes are grouped and reordered as in [15], and then, each group of cubes is mapped directly to a reversible gate cascade. For handling cubes shared by more than one outputs, CNOT gates are used for forwarding the computed values to the shared outputs.

During the final mapping, the following heuristic optimization rules are used, based on some bit patterns appearing among the cubes:

a. {11, 1-, -1} maps to a 3-input Toffoli gate with two negative controls, followed by a NOT in the target position.
b. {00, 0-, -0} maps to a 3-input Toffoli gate with two positive controls, followed by a NOT in the target position.
c. {11, 0-, -0} maps to a 3-input Toffoli gate with two negative controls, followed by a NOT in the target position.
d. {00, 1-, -1} maps to a 3-input Toffoli gate with two positive controls, followed by a NOT in the target position.

During the experimentation, it has been found that rule (a) above is mostly responsible for reduction in quantum cost for most of the benchmarks.

Example 2 An illustrative example for mapping a set of ESOP cubes into a reversible gate netlist is shown in Fig. 3. Here, the 4th, 5th, and 6th gate can be mapped using the mapping rule presented in *IV(B)*. Consider line b and c, here, the heuristic optimization rule (a) can be applied and we get reduction in terms of both gate count and quantum cost.

□

5 Experimental Results

The integrated synthesis tool has been implemented in *C* and run on an Intel dual-core-based desktop with 2.8-GHz clock and 4-GB main memory. The tool incorporates all the modules shown in Fig. 2 except the template matching module. As benchmarks, we have used various functions from the LGSynth package and Revlib [21].

In the experiment, the benchmarks in.pla format are first fed to the Exorcism-4 tool to generate the initial ESOP cube lists. Then, the proposed SA-based cube transformation tool is run to generate a new set of ESOP cubes, which are then reordered based on output sharing and finally mapped to reversible logic gates. During the gate mapping process, the rule-based optimization as explained in section IVB is also applied. Table 1 shows the results of synthesis. The first three

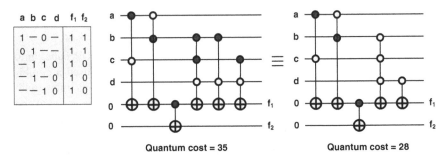

Fig. 3 Illustration of cube-to-gate mapping

Table 1 Synthesis results with benchmarks

Benchmark			Proposed approach		
Name	PIs	POs	GC	QC	Time (in s)
5xp1	7	10	79	807	0
9symm1	9	1	52	3406	0
adr4	8	5	48	652	0
alu2	10	6	86	3679	0
alu3	10	8	83	1919	0
alu4	14	8	614	38635	7
apex2	39	3	1904	376757	217
apex5	117	88	664	33803	25
apla	10	12	98	1709	0
bw	5	28	442	790	0
clip	9	5	169	3218	0
cordic	23	2	2314	111955	32
cu	14	11	27	780	0
dc2	8	7	69	1099	0
dk17	10	11	36	1013	0
duke2	22	29	206	6165	1
e64	65	65	87	24345	1
f51 m	14	8	418	25119	5
frg2	143	139	1989	114239	271
misex1	8	7	63	352	0
misex3	14	14	1596	54132	10
rd84	8	4	81	1965	0
root	8	5	93	1583	0
spla	16	46	1008	45478	6
sym10	10	1	83	5990	1
table 3	14	14	995	32286	3
table 5	17	15	1003	28139	2
vg2	25	8	222	17830	2

Table 2 Comparison with other works in terms of quantum cost

Benchmark	Method 1 [6]	Method 2 [17]	Method 3 [20]	Method 4 [15]	Method 5 [16]	Proposed
5xp1	865	693	695	786	1349	807
9symm1	16487	4261	–	10943	5781	3406*
adr4	–	618	–	–	770	652
alu2	4476	3953	–	–	5215	3679*
alu3	–	2229	–	–	–	1919*
alu4	43850	34534	–	41127	48778	38635
apex2	–	595594	–	–	–	376757*
apex5	–	40835	36221	33830	–	33803*
apla	–	2879	–	1683	–	1709
bw	–	3106	1410	637	–	790
clip	4484	3059	–	3824	6616	3218
cordic	–	–	–	187620	349522	111955*
cu	–	–	–	781	1332	780*
dc2	–	–	–	1084	1956	1099
dk17	–	–	–	1014	1976	1013*
duke2	10456	–	–	–	–	6165*
e64	–	–	33591	–	–	24345*
f51 m	–	–	–	28382	34244	25119*
frg2	–	–	114976	112008	–	114239
misex1	466	–	–	332	1017	352
misex3	67206	–	–	49076	122557	54132
rd84	2062	–	–	–	2598	1965*
root	–	–	–	1811	3486	1583*
spla	49419	–	–	–	–	45478*
sym10	35227	–	–	–	9717	5990*
table 3	35807	–	–	–	86173	32286*
table 5	34254	–	–	–	–	28139*
vg2	18417	–	–	–	–	17830*

columns of the table specify the benchmark name, and the number of primary inputs and outputs. The next three columns specify the gate count (GC), quantum cost (QC), and the run time in seconds.

Table 2 compares the synthesis results with five recent methods [6, 15–17, 20]. It may be noted that the results of [17] that have been shown in the table are without template matching, which has not been used by the other four methods, and are expected to improve the quantum costs for all the methods. Also since all the papers have not shown their results for the same set of benchmarks, some entries in the table are shown as '-' (result not available).

It can be seen from Table 2 that for 18 out of the 28 benchmarks reported, the proposed approach gives the lowest quantum cost (highlighted and marked by asterisks) among all reported works. Another advantage of the proposed method is that the synthesis time is very less, with the largest circuit (*frg2*) taking <5 min. The following points can be noted from the comparison table.

- Out of the 10 benchmarks for which the proposed method does not give the best results, 5 of them are small circuits with ($PI + PO < 20$), and also the differences in quantum cost are not so high. For 4 of the other benchmarks (alu4, apla, Fig. 2, and misex3), the differences in quantum cost is <10 %. For the benchmark *bw*, however, the difference is about 19 %. This is because for this function, $PI = 5$ and $PO = 28$, and there are a large number of cubes sharing various outputs for which Method 4 [15] is best suited.
- As compared to [6], the proposed method gives better results for all the 14 benchmarks compared with.
- As compared to [17], it may be noted that for most of the benchmarks where ($PI + PO < 20$), [17] gives better results compared to the proposed approach, but for larger circuits (like *apex2*, *apex4*, *apex5*), our method gives far better results. Overall, out of 11 benchmarks, our method provides better results for 7.
- Similar trend is observed for [20], where out of 5 benchmarks compared with, they have got better results for one where ($PI + PO < 20$), while for the larger benchmarks the proposed method gives better results.
- As compared to [15], among the 16 compared benchmarks, our results are better for 9 of them. But here for two benchmarks (*9symm1* and *cordic*), we have got significant reductions (69 and 40 %) as compared to their method. This is due of the fact that for these two benchmarks, the number of outputs is 1 and 2, respectively, and so shared cubes are very less (or nonexistent).
- And finally, as compared to [16], the proposed method gives significantly better results for all of the 17 benchmarks compared with.

It may be noted that the proposed method is more general as compared to similar published works and considers optimization of ESOP cubelist as well as that during gate netlist generation. As a future work, the power of the tool as implemented can be enhanced by incorporating additional rules.

6 Conclusion

An approach to synthesis of reversible logic circuits has been reported in this paper, which is based on ESOP realization of the function, and selective application of rules for optimization. The approach is shown to give better quantum costs compared to the best published results for many of the benchmarks used for comparison. For some of the benchmarks, however, the results are found to be worse since the rules used could not be applied to advantage on the list of cubes generated by Exorcism-4, which is more tuned to minimizing the number of cubes and not the quantum cost. As a future work, it is planned to implement an indigenous scheme for generating the initial set of ESOP cubes targeting quantum cost as the metric to be minimized, and better cube clustering and gate mapping techniques specifically targeting negative-control Toffoli gates. Also, template-based optimizations for both positive- and negative-control Toffoli gates shall be implemented to reduce the cost of implementation.

References

1. Barenco, A., Bennett, H.H., Cleve, R., DiVinchenzo, D.P., Margolus, N., Shor, P., Sleator, T., Smolin, J.A., Weinfurter, H.: Elementary gates for quantum computation. Phys. Rev. A (At. Mol. Opt. Phy.) **52**(5), 3457–3467 (1995)
2. Bennett, C.H.: Logical reversibility of computation. J. IBM Res. Dev. **17**, 525–532 (1973)
3. Bèrut, A., Arakelyan, A., Petrosyan, A., Ciliberto, S., Dillenschneider, R., Lutz, E.: Experimental verification of Landauer's principle linking information and thermodynamics. Nature **483**(3), 187–189 (2012)
4. Datta, K., Rathi, G., Sengupta, I., Rahaman, H.: Synthesis of reversible circuits using heuristic search method. In: Proceedings of 25th International Conference on VLSI Design, pp. 328–333 (2012)
5. Datta, K., Shrivastav, V., Sengupta, I., Rahaman, H.: Reversible logic implementation of AES algorithm. In: Proceedings of Design and Technology of Integrated Systems (DTIS), March 2013
6. Drechsler, R., Finder, A., Wille, R.: Improving ESOP-based synthesis of reversible logic using evolutionary algorithms. In: Proceedings of International Conference on Applications of Evolutionary Computation (Part II), pp. 151–161 (2011)
7. Fazel, K., Thornton, MA., Rice, J.: ESOP-based Toffoli gate cascade generation. In: Proceedings of IEEE Pacific Rim Conference on Communications, Computers and Signal Processing, pp. 206–209 (2007)
8. Finder, A., Drechsler, R.: An evolutionary algorithms for optimization of pseudo Kronecker expressions. In: Proceedings of International Symposium on Multi-Valued Logic, pp. 150–155 (2010)
9. Grosse, D., Wille, R., Dueck, G.W., Drechsler, R.: Exact multiple control Toffoli network synthesis with SAT techniques. IEEE Trans. CAD Integr. Circuits Syst. **28**(5), 703–715 (2009)
10. Gupta, P., Agrawal, A., Jha, N.K.: An algorithm for synthesis of reversible logic circuits. IEEE Trans. CAD Integr. Circuits Syst. **25**(11), 2317–2329 (2006)
11. Landauer, R.: Irreversibility and heat generation in computing process. J. IBM Res. Dev. **5**, 183–191 (1961)
12. Maslov, D., Dueck, G.W.: Quantum circuit simplification and level compaction. IEEE Trans. CAD Integr. Circuits Syst. **27**(3), 436–444 (2008)
13. Mishchenko, A., Perkowski, M.: Fast heuristic minimization of exclusive-sums-of-products. In: Proceedings of 6th Reed-Muller Workshop, pp. 242–250 (2001)
14. Moore, G.E.: Cramming more components onto integrated circuits. J. Electron. **38**(8), 183–191 (1965)
15. Nayeem, N., Rice, J.E.: A shared-cube approach to ESOP-based synthesis of reversible logic. Facta Universitatis of NiÊ, Elec Energ. **24**(3), 385–402 (2011)
16. Rice, J., Fazel, K., Thornton, M., Kent, K.: Toffoli gate cascade generation using ESOP minimization and QMDD-based swapping. In: Proceedings of 14th Reed-Muller Workshop, pp. 63–72 (2009)
17. Rice, J.E., Nayeem, N.: Ordering techniques for ESOP-based Toffoli cascade generation. In: Proceedings of IEEE Pacific Rim Conference on Communications, Computers and Signal Processing (PACRIM), pp. 274–279 (2011)

18. Rice, J.E., Suen, V.: Using autocorrelation coefficient-based cost functions in ESOP-based Toffoli gate cascade generation. In: Proceedings of 23rd Canadian Conference on Electrical and Computer Engineering (CCECE), pp. 1–6 (2010)
19. Sanaee, Y., Dueck, GW.: Generating Toffoli networks from ESOP expressions. In: Proceedings of IEEE Pacific Rim Conference on Communications, Computers and Signal Processing (PACRIM), pp. 715–719 (2009)
20. Sanaee, Y., Dueck, G.W.: ESOP-based Toffoli network generation with transformations. In: Proceedings of 40th International Symposium on Multiple-Valued Logic, pp. 276–281 (2010)
21. Soeken, M., Frehse, S., Wille, R., Drechsler, R.: Revkit: a toolkit for reversible circuit design. In: Proceedings of Workshop on Reversible Computation. Revkit is available at http://www.revkit.org (2010)
22. Wille,R., Drechsler, R.: BDD-based synthesis of reversible logic for large functions. In: Proceedings of Design Automation Conference, pp. 270–275 (2009)
23. Wille, R., Drechsler, R., Oswald, C., Garcia-Ortiz, A.: Automatic design of low-power encoders using reversible circuit synthesis. In: Proceedings of Design Automation Test in Europe (DATE), pp. 208–212 (2012)

An Efficient Algorithm for Reducing Wire Length in Three-Layer Channel Routing

Swagata Saha Sau and Rajat Kumar Pal

Abstract In VLSI physical design automation, channel routing problem (CRP) for minimizing total wire length to interconnect the nets of different circuit blocks is one of the most challenging requirements to enhance the performance of a chip to be designed. Interconnection with minimum wire length occupies minimum area and has minimum overall capacitance and resistance present in a circuit. Reducing the total wire length for interconnection minimizes the cost of physical wire segments required, signal propagation delays, electrical hazards, power consumption, the chip environment temperature, the heat of the neighboring interconnects or transistors, and the thermal conductivity of the surrounding materials. Thus, it meets the needs of green computing and has a direct impact on daily life and environment. Since the problem of computing minimum wire length routing solutions for three-layer no-dogleg general channel instance is NP-hard, it is interesting to develop heuristic algorithms that compute reduced total wire length routing solutions within practical time limit. In this paper, we have developed an efficient polynomial time graph-based heuristic algorithm that minimizes the total wire length for most of the benchmark channel instances available in the reserved three-layer no-dogleg Manhattan channel routing model. The results we compute are highly encouraging in terms of efficiency and performance of our algorithm in comparison to other existing algorithms for computing the same.

Keywords Channel routing problem · Wire length minimization · Manhattan routing · No-doglegging · Parametric difference

S.S. Sau (✉)
Department of Computer Science, Sammilani Mahavidyalaya, Baghajatin, Kolkata, West Bengal 700075, India
e-mail: swagatasahasau@gmail.com

R.K. Pal
Department of Computer Science and Engineering, University of Calcutta, Kolkata, West Bengal 700009, India
e-mail: pal.rajatk@gmail.com

© Springer India 2015
R. Chaki et al. (eds.), *Applied Computation and Security Systems*, Advances in Intelligent Systems and Computing 305, DOI 10.1007/978-81-322-1988-0_9

1 Introduction

The channel routing problem (CRP) for minimizing total wire length to interconnect the nets of different circuit blocks is one of the most challenging problems
in VLSI physical design automation. Reducing the total wire length for interconnection not only minimizes the total amount of area required, the cost of
physical wire segments needed, signal propagation delays, electrical hazards,
power consumption, the chip environment temperature, and the heat of the
neighboring interconnects or transistors, but also enhances the performance of the
chip to be designed and protect the environment [7]. Thus, it meets the requirements of green computing and has a direct impact on daily life and environment.

As the routing areas are fixed in gate arrays, hence minimizing total wire length
is one of the most challenging tasks in gate array design. Even in the case of
custom design and standard cell chip design, wire length minimization is one of
the usual objective functions as well as important from high-performance
requirement point of view, where routing space could be adjusted. CRP for wire
length minimization is an NP-hard problem [8, 13]. So it is interesting to develop a
heuristic algorithm that computes reduced total wire length routing solutions
within a reasonable time limit. There are several algorithms developed for
reducing area as well as total wire length for VLSI channel routing [1–3, 5–14]. In
this section, we shortly review them below.

Routing is a process of interconnecting the terminals (pins) on the periphery of
different cells or blocks using wire segments on different layers of interconnect.
Routing within a chip is obtained usually in two phases: in global routing phase
and in detailed routing phase. In the global routing phase, the router identifies the
empty regions through which a particular interconnection will be made, and in the
detailed routing phase, actual interconnections are realized within the rectangular
region of each channel and switchbox. A switchbox has fixed terminals on three or
four sides. Channel is a rectangular routing region with two open ends (left and
right), and the top and bottom sides of a channel have two rows of fixed terminals
(top terminals and bottom terminals). Terminals are aligned vertically at columns
and assigned by some integer numbers: zero for vacant terminals, not to be connected. A set of terminals that need to be electrically connected together is called a
net and assigned by the same integer number. Channel is specified by channel
specification or net list. The channel length is the total number of pins in the top
(or bottom) row of a channel, and the height of the channel depends on the number
of tracks that are needed to route all the nets within the channel.

A net containing only two terminals is called two terminal net, and a net with
more than two terminals is called multi-terminal net. For floating terminals, the left
and/or the right end of the nets is not fixed. The set of nets that enter into the
channel from left (right) form the left (right) connection set and represented as
LCS (RCS) in the channel specification. The distance between the leftmost and
rightmost column positions of a net n_i is known as interval I_i or span of net n_i. For
reserved layer Manhattan routing model, only horizontal and vertical wire

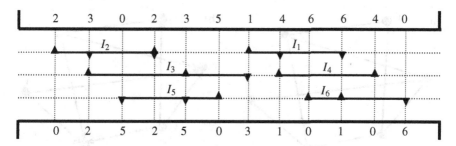

Fig. 1 The channel instance *RKPC2* with six nets

segments are used for interconnection and placed in the respective layer(s). In case of no-dogleg routing, the horizontal wire segment of a net is not split into two or more parts and assigned to different tracks. Figure 1 shows the channel instance *RKPC2* with six nets [8].

Wire length minimization of the CRP means to obtain all the interconnections obeying two constraints—horizontal constraints and vertical constraints of the channel. A horizontal constraint exists between two nets if their horizontal segments overlap, when they are placed on the same track. Horizontal constraints of the CRP can be represented by an undirected graph, known as horizontal constraint graph, HCG (V, E_h), where $V = \{v_i | v_i$ represents interval I_i corresponding to net $n_i\}$ and $E_h = \{(v_i, v_j) | I_i$ and I_j overlap$\}$ [8–11]. A vertical constraint exists between two nets if they have terminals in the same column, and vertical constraints determine the order in which the intervals or nets should be assigned from top to bottom across the channel height. Vertical constraints can be represented by a directed graph, known as vertical constraint graph (VCG), VCG (V, E_v), where $V = \{v_i | v_i$ represents interval I_i corresponding to net $n_i\}$ and $E_v = \{(v_i, v_j) | n_i$ has vertical constraint with n_j in some column of the channel$\}$ [4, 9–12]. Although a vertical constraint implies a horizontal constraint, the reverse is not necessarily true.

The local density of a column is the maximum number of nets passing through the column, and density of a channel (d_{max}) is the maximum of all the local densities. v_{max} is the number of vertices belonging to a longest path of an acyclic

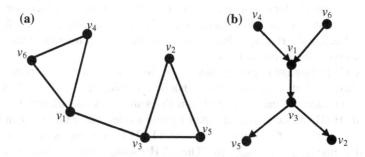

Fig. 2 **a** The HCG of the channel instance in Fig. 1. **b** The VCG of the channel instance in Fig. 1

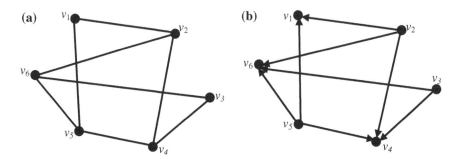

Fig. 3 **a** The HNCG of the channel instance in Fig. 1. **b** The OHNCG of the channel instance in Fig. 1

VCG. In Fig. 2a, b, the HCG and the VCG of the channel instance *RKPC2* are shown, respectively. The channel density $d_{\max} = 3$ and $v_{\max} = 4$ of the channel instance *RKPC2*.

In this paper, we represent horizontal constraints using the complement of HCG, known as the horizontal non-constraint graph (HNCG), HNCG (V, E'_h), where $V = \{v_i \mid v_i$ represents interval I_i corresponding to net $n_i\}$ and $E'_h = \{(v_i, v_j) \mid (v_i, v_j) \notin E_h\}$. An edge between a pair of vertices of an HNCG represents that their corresponding intervals are non-overlapping. We consider VCG and HNCG to compute a set of non-overlapping intervals for each track. We provide transitive orientation to each edge (v_i, v_j) of the HNCG according to the position of the corresponding intervals from left (right) to right (left) in a channel and obtain an oriented horizontal non-constraint graph (OHNCG). Figure 3a, b show HNCG and OHNCG of the channel instance *RKPC2*, respectively.

In three-layer routing, wires can be assigned to layers in two ways in the reserved layer routing model. In the first model, a vertical layer is flanked by two horizontal layers (i.e., HVH), and in the second model, a horizontal layer is flanked by two vertical layers (i.e., VHV). In the case of VHV Manhattan routing model, vertical constraints are ineffective as the top terminal and bottom terminal of different nets belonging to a column can be connected through wires placing in separate vertical layers. So in this case, we can apply *Minimum_Clique_Cover_1* (*MCC1*) algorithm along with parametric difference of the nets incorporated in horizontal constraint graph to compute a set of non-overlapping intervals for each track [8]. Each of the routing solutions requires exactly $d_{\max}$ number of tracks, and it can be computed in polynomial time.

In the HVH Manhattan routing model, two horizontal layers are separated by a vertical layer, so at most two sets of non-overlapping intervals which are not vertically constrained can be placed into the same track of different horizontal layers. In HVH (no-dogleg) Manhattan routing model, the minimum number of tracks required, i.e., the trivial lower bound (TLB) on the number of tracks required is $\max(\lceil d_{\max}/2 \rceil, v_{\max})$ [8]. The HVH routing model is preferred, when $v_{\max} < \lceil d_{\max}/2 \rceil$. In this paper, we have developed an efficient high-performance algorithm *Minimum_Wire_Length_HVH* for reducing the total (vertical) wire length of a channel under the reserved three-layer Manhattan no-dogleg routing model in VLSI physical design automation and that completes 100 % routing interconnection for all the instances under consideration.

This paper is organized as follows. In Sect. 2, we formulate the problem and develop the algorithm. In Sect. 3, we include experimental results and performance of our algorithm. The paper is concluded with few remarks in Sect. 4.

2 Formulation of the Problem and the Proposed Algorithm

In this section, we state the nature of wire length minimization problem in VLSI physical design automation. Reserved three-layer no-dogleg CRP for wire length minimization is NP-hard problem [8, 13]. So development of polynomial time heuristic algorithm with minimum wire length for interconnection is truly interesting. Most of the algorithms have been developed for area minimization of a channel mainly in different routing models [2, 8], and only a few have been concentrated on wire length minimization in the unreserved unrestricted channel routing models [3, 5, 8, 10, 11]. The reducing of area usually reduces total (vertical) wire length and vice versa, but there are examples of channels in VLSI circuits where minimizing area does not minimize total wire length and vice versa [8].

Saha Sau et al. [11] have recently developed a purely graph-based polynomial time heuristic algorithm *Minimum_Wire_Length_of_CRP* for computing minimum wire length routing solutions in the reserved two-layer no-dogleg Manhattan routing model for general instances of channel specification. This algorithm is used for computing routing solutions using optimal or near optimal wire length for most well-known benchmark channels. Reducing the total wire length of a channel means reducing the total horizontal and total vertical wire length. As the terminal positions are fixed in a channel, so reducing total wire length means reducing the total vertical wire length only.

In this paper, we develop a graph-based algorithm *Minimum_Wire_Length_HVH* to reduce the total (vertical) wire length of feasible channel routing solutions in the reserved three-layer no-dogleg Manhattan routing model. If the total number of the top terminals TT_i of net n_i is more than the total number of bottom terminals BT_i of the same net, then the assignment of net n_i toward the top row reduces the total vertical wire length and vice versa. Difference between TT_i and BT_i is called *parametric difference* of net n_i and it is denoted by pd_i.

Our proposed algorithm *Minimum_Wire_Length_HVH* on three-layer no-dogleg HVH channel routing for minimizing total (vertical) wire length is based on the algorithm *Modified_MCC1*, which has been developed in the two-layer no-dogleg Manhattan routing model for doing the same [11]. The input to the algorithm is channel specification or net list of a channel, and the output is a feasible three-layer channel routing solution with minimum total wire length. In our proposed algorithm, we assign weights ⟨height ht_i, parametric difference pd_i, position p_i of the net n_i from the left of the channel, interval $span_i$⟩ to the vertices in OHNCG in the form of a 4-tuple and use the algorithm *Maximum_Weighted_Clique* (*MWC*) of Golumbic for selecting a desired set of nets with non-overlapping spans or for computing a clique in the comparability graph OHNCG [4].

To compute a clique, we maximize all the parameters of the 4-tuple except one parameter, i.e., position p_i of the net n_i from the left of the channel. In each iteration, we compute a set of non-overlapping intervals or clique C_t^1 of OHNCG from the set of source vertices S_1 of the current VCG using *MWC* of Golumbic and assign its corresponding intervals to the tth track of the first horizontal layer. If C_t^1 $\in S_1$, then we compute another set of non-overlapping intervals or clique C_t^2 of OHNCG from the remaining set $S_1 - C_t^1$ of source vertices of the current VCG using *MWC* of Golumbic, assign its corresponding intervals to the tth track of the second horizontal layer, and process for the next iteration, if a net is yet to assign. If $C_t^1 = S_1$, then no other intervals are left to assign to the tth track of the second horizontal layer. In that case, the tth track of the second horizontal layer remains vacant and then processes for the next iteration. The algorithm *Minimum_Wire_Length_of_CRP* for computing reduced wire length channel routing solutions in the reserved two-layer no-dogleg Manhattan model assigns the non-overlapping sets of intervals to tracks using the sandwich method, once for the top track, then for a bottom track, and so on, but we divide our way of implementation into four sub-modules and execute each of them, and finally take the optimal one among the routing solutions computed using all these sub-modules. Here, we state the four sub-modules as follows.

Module 1: *Track assignment from top to bottom and scan of a channel from left to right*: In each iteration, we scan a channel from the left end to the right end of the channel and assign a set of non-overlapping intervals to the available topmost track under certain constraints.

Module 2: *Track assignment from bottom to top and scan of a channel from left to right*: In each iteration, we scan a channel from the left end to the right end of the channel and assign a set of non-overlapping intervals to the available bottommost track under certain constraints.

Module 3: *Track assignment by top–bottom sandwiched and scan of a channel from left to right*: We scan a channel from the left end to the right end of the channel, and in each odd iteration assign a set of non-overlapping intervals to the available topmost track and in each even iteration assign a set of non-overlapping intervals to the available bottommost track under certain constraints.

Module 4: *Track assignment by bottom–top sandwiched and scan of a channel from left to right*: We scan a channel from the left end to the right end of the channel, and in each odd iteration assign a set of non-overlapping intervals to the available bottommost track and in each even iteration assign a set of non-overlapping intervals to the available topmost track under certain constraints.

As these four modules are independent to each other on computational aspect, hence these modules may be computed in parallel. Channel specification of a channel instance is the input of our proposed algorithm, and output is a feasible three-layer channel routing solution with minimum total wire length. Let us

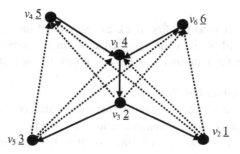

Fig. 4 The hybrid graph structure combining the graphs in Figs. 2b and 3b of the channel instance *RKPC2* shown in Fig. 1

consider the channel instance *RKPC2* and we discuss our proposed algorithm considering the Module 1, i.e., *top to bottom track assignment and left to right scan of a channel* in three-layer (HVH) no-dogleg routing model.

Let us consider an input channel specification as follows:

$$
\begin{array}{cccccccccccc}
2 & 3 & 0 & 2 & 3 & 5 & 1 & 4 & 6 & 6 & 4 & 0 \\
0 & 2 & 5 & 2 & 5 & 0 & 3 & 1 & 0 & 1 & 0 & 6
\end{array}
$$

First, we construct a *net_information_list* containing all the nets scanning from the left of the channel. Construct HNCG (V, E'_h) and compute its transitively oriented graph OHNCG (V, F) based on the natural transitive orientation from left to right of the channel. Construct VCG (V, E_v) and compute the maximum height of each vertex from sink vertices and represent it as the height ht_i of vertex v_i for net n_i. Combine OHNCG (with orientation) and VCG and obtain a hybrid graph structure where vertices represent different intervals (or nets) of the given channel instance, where edges of OHNCG and VCG are differentiated by some means. Assign natural numbers starting from 1 through n to the vertices of the hybrid graph structure according to their starting column position from left to right of the channel. Figure 4 shows the hybrid graph structure combining the graphs in Figs. 2b and 3b of the channel instance *RKPC2*. We assign the 4-tuple weight to each vertex v_i of this graph as ⟨Height ht_i, parametric difference pd_i, position p_i of the net n_i from the left of the channel, and interval $span_i$⟩ in the hybrid graph structure.

Now the *net_information_list* = {2, 3, 5, 1, 4, 6}. In subsequent iterations, we compute the following for the channel instance *RKPC2* shown in Fig. 1.

Iteration 1:

S = Set of source vertices of hybrid graph structure according to the current VCG = {4, 6}.

Here, $ht_4 = ht_6 = 4$, but $pd_4 > pd_6$, hence net 4 is selected first. As net 4 and net 6 are overlapping and there is no other net(s) in S, hence clique $C^1_1 = \{4\}$.

Assign net 4 to the first track of the first horizontal layer and find another clique from $S - C^1_1$.

Clearly, $C_1^2 = \{6\}$.

So, we assign net 6 to the first track of the second horizontal layer. Delete vertices 4 and 6 and all its connecting edges from the hybrid graph structure. Delete 4 and 6 from *net_information_list* and go for the next iteration.

Iteration 2:

Now the *net_information_list* = $\{2, 3, 5, 1\}$.

S = Set of source vertices of hybrid graph structure according to the current VCG = $\{1\}$. Similarly, we can compute clique C_2^1 as $\{1\}$.

We assign net 1 to the second track of the first horizontal layer and as $S - C_2^1$ = NULL, hence nothing to be assigned to the second track of the second horizontal layer.

We delete vertex 1 and all its connecting edges from the hybrid graph structure and also delete 1 from *net_information_list* and go for the next iteration.

Iteration 3:

At the beginning of this iteration, *net_information_list* = $\{2, 3, 5\}$.

S = Set of source vertices of hybrid graph structure according to the current VCG = $\{3\}$.

We compute clique C_3^1 as $\{3\}$ for the third track of the first horizontal layer.

Assign net 3 to the third track of the first horizontal layer and as $S - C_3^1$ = NULL, hence nothing to be assigned to the third track of the second horizontal layer.

We delete vertex 3 and all its connecting edges from the hybrid graph structure and delete 3 from *net_information_list* and go for next iteration.

Iteration 4:

Now, *net_information_list* = $\{2, 5\}$.

S = Set of source vertices of hybrid graph structure according to the current VCG = $\{2, 5\}$.

Here, $ht_2 = ht_5 = 1$, but $pd_2 > pd_5$, hence net 2 is selected first. Now as the nets 2 and 5 overlap in the channel, hence clique $C_4^1 = \{2\}$.

We assign net 2 to the fourth track of the first horizontal layer and find another clique from $S - C_4^1$.

Clearly, the clique C_4^2 is $\{5\}$ as it also belongs to S.

So, we assign net 5 to the fourth track of the second horizontal layer. Then, we delete vertices 2 and 5 and all their connecting edges from the hybrid graph structure. Then, also we delete 2 and 5 from *net_information_list*.

As *net_information_list* is NULL, hence the algorithm terminates. Figure 5 shows the reduced wire length routing solution of the channel instance *RKPC2* using the algorithm developed in this paper. This routing solution requires four tracks and total vertical wire length of 40 units.

Now, we analyze the time complexity of the algorithm *Minimum_Wire_Length _HVH* devised in this paper. The algorithm has an initial part of computation as well as a part of iterative computation. The time complexity to construct the HNCG is $O(n + e)$, and the time complexity to construct the VCG is $O(n)$ time, where n is the number of nets in the given channel and e is the size of HNCG. The time

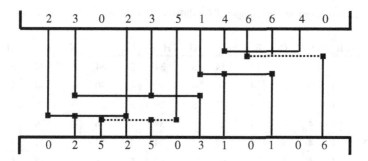

Fig. 5 The reduced wire length routing solution of the channel instance *RKPC2* computed using algorithm developed in this paper, where total number of tracks = 4 and the total *vertical* wire length = 40 units

complexity of each iteration is also $O(n + e)$. If the number of iterations required to route a channel is t, then the time complexity of the iterative part of the algorithm is $O(t(n + e))$, where n is the number of nets in the given channel, e is the size of HNCG, and t is the number of tracks required to route the channel. Hence, the overall time complexity of our algorithm is $O(t(n + e))$, where the variables concerned are as stated above. The algorithm *Minimum_Wire_Length_HVH* correctly computes a routing solution in the reserved three-layer no-dogleg CRP for a general channel instance (where the VCG does not contain any cyclic vertical constraint). We state this result in the form of the following theorem.

Theorem 1 *Algorithm Minimum_Wire_Length_HVH successfully computes a three-layer HVH routing solution in the reserved no-dogleg Manhattan channel routing model for a general channel specification without any cyclic vertical constraint in the VCG and takes time $O(t(n + e))$, where n is the number of nets belonging to the channel, e is the size of the HNCG, and t is the number of iterations or the number of tracks required to route the given channel.*

3 Experimental Results and Performance of Our Algorithm

In this section, we show some experimental results that are computed using the algorithm *Minimum_Wire_Length_HVH* developed in Sect. 2. Results are shown in Table 1 for the benchmark instances available in [8, 14]. Table 1 includes different values relating a number of benchmark channel instances as A associating other parameters like B: Channel density (d_{max}), C: Length of the longest path in VCG (v_{max}), D: Trivial lower bound on the number of tracks in the HVH model, E: Number of tracks computed using the developed algorithm for Module 1, F: Total (vertical) wire length computed using the algorithm for Module 1, G: Number of

Table 1 Experimental results computed for different channel instances available in literature [8, 14]

A	B	C	D	E	F	G	H	I	J	K	L	M	N
RKPC1	3	3	3	3	34	3	34	3	34	3	34	3	34
RKPC2	3	4	4	4	40	4	40	4	40	4	40	4	40
RKPC3	4	3	3	3	27	3	26	3	26	3	26	3	26
RKPC4	4	4	4	4	29	4	34	4	26	4	26	4	26
RKPC5	4	4	4	4	37	4	34	4	27	4	31	4	27
RKPC6	4	5	5	5	90	5	89	5	85	5	85	5	85
RKPC7	4	3	3	5	59	5	57	5	57	5	54	5	54
RKPC8	5	5	5	5	78	5	83	5	79	5	79	5	78
RKPC9	6	5	5	6	173	6	169	6	174	6	164	6	164
Ex. 1	12	7	7	7	164	7	168	7	154	7	154	7	154
Ex. 2	15	4	8	8	223	9	306	9	254	9	264	8	223
Ex. 3(a)	15	4	8	10	349	9	377	9	315	9	310	9	310

where A Channel instance, B channel density (d_{max}), C length of the longest path in VCG (v_{max}), D Trivial lower bound in HVH, E number of tracks computed using the algorithm for module 1, F total (vertical) wire length computed using the algorithm for module 1, G number of tracks computed using the algorithm for module 2, H total (vertical) wire length computed using the algorithm for module 2, I number of tracks computed using the algorithm for module 3, J total (vertical) wire length computed using the algorithm for module 3, K number of tracks computed using the algorithm for module 4, L total (vertical) wire length computed using the algorithm for module 4, M optimal number of tracks computed using the devised algorithm, and N optimal total (vertical) wire length computed using the algorithm

tracks computed using the algorithm for Module 2, H: Total (vertical) wire length computed using the algorithm for Module 2, I: Number of tracks computed using the algorithm for Module 3, J: Total (vertical) wire length computed using the algorithm for Module 3, K: Number of tracks computed using the algorithm for Module 4, L: Total (vertical) wire length computed using the algorithm for Module 4, M: Optimal number of tracks computed using the algorithm, and N: Optimal total (vertical) wire length computed using the algorithm.

The channel routing solutions using our algorithm *Minimum_Wire_Length_HVH* in the reserved no-dogleg three-layer HVH channel routing model drastically reduce the total (vertical) wire length in comparison to the three-layer HVH routing solutions computed using algorithm *TAH* for wire length minimization in the same model [8]. Results are shown in Table 2 for several benchmark instances available in [8, 14]. In Table 2, A: Channel instance, B: Channel density (d_{max}), C: Length of the longest path in VCG (v_{max}), D: Number of tracks required computed using algorithm *TAH* in the supposed routing model for wire length minimization in [8], E: Total (vertical) wire length computed using the algorithm *TAH* in the said routing model for wire length minimization in [8], F: Number of tracks computed using our algorithm *Minimum_Wire_Length_HVH* in the assumed routing model, and G:

Table 2 Experimental results computed for different channel instances available in literature [8, 14]

A	B	C	D	E	F	G
Ex. 1	12	7	7	156	7	154
Ex. 2	15	4	8	275	8	223
Ex. 3(a)	15	4	8	338	9	310
Ex. 3(b)	17	9	10	474	10	446
Ex. 3(c)	18	6	10	557	10	520
Ex. 4(b)	17	13	13	1,017	15	909
Ex. 5	20	3	10	728	10	557
DDE	19	23	23	3,041	14	2,988

Total (vertical) wire length computed using our algorithm *Minimum_Wire_Length_HVH* in the implicit channel routing model. Note that the devised algorithm grippingly reduces the total (vertical) wire length for all the instances we consider and include in Table 2.

4 Conclusion

In this paper, we have developed reserved three-layer no-dogleg graph-based heuristic algorithm for computing reduced wire length VLSI channel routing solutions. Three-layer CRP for minimizing wire length is an NP-hard problem [8, 13]. Here, we have developed a polynomial time heuristic algorithm for reducing total (vertical) wire length in the reserved three-layer HVH no-dogleg Manhattan channel routing model. The time complexity of the algorithm developed herein is $O(t(n + e))$, where t is the number of tracks to route the channel, n is the number of nets in the given channel instance, and e is the size of the HNCG. The performance of our algorithm is highly encouraging in terms of the results computed and included in the tables. Here, we like to include a few probable extensions of our work: (1) extension in computing reduced wire length dogleg routing solutions, (2) extension in computing reduced wire length as well as reduced area dogleg and no-dogleg routing solutions, and (3) extension in computing reduced wire length routing solutions in the reserved multilayer channel routing models.

References

1. Alpert, C.J., Mehta, D.P., Sapatnekar, S.S.: Handbook of Algorithms for Physical Design Automation. CRC Press, New York (2009)
2. Cong, J., Wong, D.F., Liu, C.L.: A new approach to the three-layer channel routing problem. In: Proceedings of IEEE ICCAD, pp. 378–381 (1987)

3. Formann, M., Wagner, D., Wagner, F.: Routing through a dense channel with minimum total wire length. In: Proceedings of 2nd Annual ACM-SIAM Symposium, pp. 475–482 (1991)
4. Golumbic, M.C.: Algorithmic Graph Theory and Perfect Graphs. Academic Press, New York (1980)
5. Hashimoto, A., Stevens, J.: Wire routing by optimizing channel assignment within large apertures. In: Proceedings of 8th ACM Design Automation Workshop, pp. 155–169 (1971)
6. Hong, C., Kim, Y.: The efficient hybrid approach to channel routing problem. Int. J. Adv. Sci. Technol. **42**, 61–68 (2012)
7. Lienig, J.: Introduction to electromigration-aware physical design (invited talk). In: Proceedings of ISPD'06, pp. 39–46 (2006)
8. Pal, R.K.: Multi-layer Channel Routing: Complexity and Algorithms, Narosa Publishing House, New Delhi (Also published from CRC Press, Boca Raton, USA and Alpha Science International Ltd., UK) (2000)
9. Pal, R.K., Datta, A.K., Pal, S.P., Das, M.M., Pal, A.: A General Graph Theoretic Framework for Multi-layer Channel Routing. In: Proceedings of 8th VSI/IEEE International Conference on VLSI Design, pp. 202–207 (1995)
10. Pal, R.K., Datta, A.K., Pal, S.P., Pal, A.: Resolving Horizontal Constraints and Minimizing Net Wire Length for Multi-layer Channel Routing. In: Proceedings of IEEE Region 10's 8th Annual International Conference on Computer, Communication, Control, and Engineering (TENCON 1993), vol. 1, pp. 569–573 (1993)
11. Sau, S.S., Pal, A., Mandal, T.N., Datta, A.K., Pal, R.K., Chaudhuri, A.: A Graph based Algorithm to Minimize Total Wire Length in VLSI Channel Routing. In: Proceedings of International 2011 IEEE Conference on Computer Science and Automation Engineering (CSAE), vol. 3, pp. 61–65 (2011)
12. Somogyi, K.A., Recski, A.: On the complexity of the channel routing problem in the dogleg-free multilayer manhattan model, ACTA Polytechnica Hungarica, vol. 1, no. 2 (2004)
13. Szymanski, T.G.: Dogleg channel routing is NP-complete. IEEE Trans. CAD Integr. Circ. Syst. **4**, 31–41 (1985)
14. Yoshimura, T., Kuh, E.S.: Efficient algorithms for channel routing, IEEE Trans. CAD Integr. Circ. Syst. **CAD-1**, 25–35 (1982)

A New Move Toward Parallel Assay Operations in a Restricted Sized Chip in Digital Microfluidics

Debasis Dhal, Arpan Chakrabarty, Piyali Datta
and Rajat Kumar Pal

Abstract Digital microfluidic biochip (DMFB) is modernizing many areas of Microelectronics, Biochemistry, and Biomedical sciences. As a substitute for laboratory experiments, it is also widely known as 'lab-on-a-chip' (LOC). Minimization in pin count and avoiding cross-contamination are some of the important design issues for realistic relevance. These days, due to urgent situation and cost efficacy, more than one assay operations are essential to be carried out at the same time. So, parallelism is inevitable in DMFB. Having an area of a given chip as a constraint, how efficiently we can use a limited sized chip and how much parallelism can be incorporated are the objectives of this paper. The paper presents a design automation flow that enhances parallelism by adopting *Connect-5* structure of pin configuration and considering cross-contamination problem as well. The algorithm developed in this paper assumes array-based partitioning of modules as pin-constrained design technique, where a constant number of pins have been used for desired scheduling of reagent and sample droplets. To avoid cross-contamination and at the same time to minimize the delay required for washing, wash droplet scheduling and proper placement of modules to minimize wash operations are also taken care of.

D. Dhal (✉)
Dum Dum Subhasnagar High School (H.S.), 43, Sarat Bose Road,
Kolkata 700 065, West Bengal, India
e-mail: todevcse@rediffmail.com

A. Chakrabarty · P. Datta · R.K. Pal
Department of Computer Science and Engineering, University of Calcutta,
92, A.P.C. Road, Kolkata 700 009, West Bengal, India
e-mail: arpan250506@gmail.com

P. Datta
e-mail: piyalidatta150888@gmail.com

R.K. Pal
e-mail: pal.rajatk@gmail.com

© Springer India 2015
R. Chaki et al. (eds.), *Applied Computation and Security Systems*, Advances in Intelligent
Systems and Computing 305, DOI 10.1007/978-81-322-1988-0_10

Keywords Lab-on-a-chip · Cross-contamination · Design automation · Sample · Reagent · Wash droplet · Pin-constrained design · Algorithm · Parallelism

1 Introduction

In recent years, there are massive revolutions in terms of performance and efficiency while using biochips to detect the status of samples. It is the most advanced device nowadays in the micro-level for diagnosing (analyzing, testing, and detecting) some specimen such as DNA, blood, saliva, stool, cough, urine, and many others that we like to examine in our everyday life. There are some challenging scopes to betterment the performance of this biochip. This device is also known as digital microfluidic biochip (DMFB) [1–5]. It can perform all the tasks of droplet creation (dispense of droplet), transportation (routing of droplet), mixing (amalgamation of sample and reagent droplets), and sensing (detection of parameters present in a sample) that are much more cost-effective and time efficient in comparison with that usually done in a pathological laboratory.

Droplet-based digital microfluidics are technologies that provide fluid-handling capability on a chip. It leads to the automation of laboratory procedures; that is why it is known as 'lab-on-a-chip' (LOC). In biochemistry and biomedical sciences, microfluidic biochip has of much importance that is realized at the level of microelectronic arrays of electrodes (or cells). These devices operate on microliter or nanoliter volume of biological samples, which are routed throughout the chip using electrowetting in a 'digital' manner under clock control on a 2D array of electrodes [6, 7]. These electrodes in a DMFB combine Electronics with Biology and integrate various bioassay operations from sample preparation to detection. The foremost objective is to minimize the time required to get the result(s) of the assay using micro- and nano-level samples and reagents, where the perfectness of the results we obtain is greatly increased.

For a biochip, the efficiency is determined by the following criteria:

- *Increasing portability*: the device is to be portable with low energy consumption,
- *Higher sample throughput*: number of samples/assays per unit time,
- *Lower the cost of instrument*: development, maintenance, and testing costs of the instrument,
- *Minimizing the cost of disposables*: defining the costs per assay (together with reagent consumption),
- *Reducing the number of parameters per sample*: number of different parameters to be analyzed per sample, i.e., number of detections required,
- *Low sample and reagent consumption*: amount of sample and/or reagents required per assay,
- *Variety of unit operations performed*: the variety of laboratory operations that can be realized such as splitting, mixing, detection, and transportation,

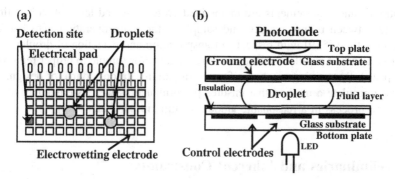

Fig. 1 **a** *Top view* of a microfluidic array with two droplets and a detection site. **b** A *side view* of digital microfluidic platform (of a cell) with a conductive glass plate present in a detection site

- *Programmability or scope of reconfiguration*: the flexibility to assay planning through software [3].

Portability, low sample and reagent consumption, and programmability are inherent qualities of a DMFB. So, we can improve some factors that are mentioned below through our work to meet the above criteria; the most important of which are to efficiently use the chips, incorporating parallelism, and to avoid cross-contamination.

Figure 1a shows a typical $n \times n$ 2D array of microfluidic biochip holding two droplets and one detection site. Figure 1b shows a side view of the biochip. It represents a typical detection site as well, where a mixed droplet can be detected optically and generates some desired results. When the LED glows and light passes through the electrode and also through the mixed droplet, the photoelectric diode measures the intensity of this light and draws some voltage against this intensity [8]. This voltage of the photoelectric diode helps to predict a set of desired outcome of the parameters present in the sample we like to test.

The concept of DMFB occurred only in two decades back. The key sense of DMFB is that a unit volume of some fluid under test is constant. It depends on the geometry of the system, an array that consisting of cells or electrodes of a matching size (to the droplet). This system is based on volume flow rate and again the volume flow rate is based on the number of droplets transported for performing some assay. This is how a droplet constitutes the fluid volume. The volume of these droplets may be several microliters.

To acquire high throughput, multiple bioassay operations are supposed to be performed concurrently. At the same time, we have to avoid droplet interference as well as contamination problems at the cost of a minimum number of pins, that is, the provision for availability of a minimum number of distinct input voltages.

To reduce the total time and to acquire accurate results in a reasonable amount of time, often the task(s) is (are) required to be performed in parallel. Hence, multiple operations often can be performed in the form of pipelining, if all the constraints and requirements are maintained up to a desired level of satisfaction. Mixing between proper reagent and sample is the main operation, which takes maximum time [5, 9] with respect to transportation and detection of droplets. So, we need to adopt parallel distribution of reagent and/or sample to proper region on a chip such that mixing can be performed in parallel. In our course of design, we like to formulate a method that ensures performance as well as efficiency of the detection process in a reasonable amount of time in parallel.

2 Preliminaries and Inherent Constraints

2.1 Preliminaries

In this section, we briefly define some of the basic terms associated to the problem of DMFB. We know that in such a chip, droplets are disposed from the outside of the array. So, there are several sources of droplets, either for sample, or for reagent, or for washing the chip.

A routing path is the passageway that a droplet uses for its movement following adjacent cells of an array through a synchronized activation and deactivation of the electrodes. This path may route from a source to a mixer, then from a mixer to a detection site, and then from there to a sink.

A mixer is a module in an array where the most important task of mixing happens. Here, different sample(s) and reagent(s) come from their respective sources and mix for detection. This mixing operation takes a maximum amount of time needed for an assay. A mixer can also be used for splitting of a droplet or dilute a sample droplet. A typical mixer takes 1,000–2,000 clock cycles.

Detection site is a small module usually formed by a single cell in the array that helps to detect the parameters present in a sample to be detected. Generally, it is done on mixed droplets, but it may often be required to detect a sample or reagent before mixing as well. As optical detection is done in such a site, the electrodes used in that cell are transparent and light of an LED can pass through it, and a photodiode placed on the top can measure the intensity of the light that can detect anomalies, if any in the sample. Usually, the number of detection sites is not many (as it is a costly module) and their locations are also tentatively fixed.

An assay is a whole operation that includes creation of droplets, their routing, mixing, and detection (of a sample's state). We usually deploy an array of electrodes that are activated and deactivated in a preferred synchronized fashion, and all subsequent steps of an assay are tracked to meet the objective(s) affirmed by the assay.

2.2 Constraints in Performing Bioassay Operations

A bioassay operation involves several tasks such as routing of droplets, mixing of droplets, detecting some parameters present in a sample, and many others. Naturally, some problem-related constraints are there; some of which are fluidic constraint, electrode constraint, time constraint, and area constraint, as briefly discussed below.

Fluidic constraint: During droplet routing, in static condition, at least one cell is supposed to be kept in between two electrodes containing two droplets to prevent unintended mixing. During movement of droplets following a particular direction, we may observe that at least a gap to two electrodes is must to avoid unwanted mixing. Hence, static and dynamic fluidic constraints [10, 11] are introduced, as these are necessary for a pair of droplets for their minimum separation on a bioassay.

Electrode constraint: In case of pin-constrained design, more than one electrodes are controlled by a single pin. This may introduce unwanted effect of voltage on some electrode, and as a result, this electrode may activate a droplet staying in an adjacent electrode inadvertently. Hence, the droplets may not move following a given schedule. This imposes several constraints during routing. If we can make proper voltage assignment over the pins, truthful movement of droplets can be guaranteed.

Timing constraint: Timing constraint in droplet routing is given by an upper bound on droplet transportation time. It is defined to have the proper synchronization among all the bioassay operations held in different modules. All the operations are pre-scheduled, and the result should be out within some specified time limit. So, there is an upper bound on time for each individual operation, which is referred to as the timing constraint.

Area constraint: We want to perform all the bioassays in a minimum chip area in view of all the above-mentioned constraints. All kinds of assignments include droplet transportation from the source of droplet to the mixing region and also to the detection site. A mixing region is supposed to be located in a proper position for utilization of total array area. So, a design must support how efficiently a chip of some fixed area can be utilized. Though we are supposed to satisfy all the constraints in isolation, maintaining all the constraints for some bioassay may introduce the problem of cross-contamination.

Cross-contamination problem: Cross-contamination occurs when the residue of one droplet transfers to another droplet with undesirable consequences, such as misleading assay outcomes, that is, incorrect diagnosis. The problem of cross-contamination may also occur when a common path is shared by two distinct droplets by fulfilling their timing constraint.

Sequencing graph: The vertices represent the assay operations (dispensing, mixing, detection, etc.), and the edges represent their mutual dependencies. This method allows the user to describe bioassay at a high level of abstraction, and it automatically maps behavioral description to the underlying microfluidic array.

2.3 Various Fundamental Operations on DMFB

The droplet formation, that is, initial metering, is the elementary unit operation of the platform. In this procedure, a proper sized droplet is created considering the size of the chip, which is predefined. Droplets can be produced from an on-chip reservoir in three steps [5]. First, a liquid column is extruded from the reservoir by activating a series of adjacent electrodes. Second, once the column overlaps the electrode on which the droplet is to be formed, all the remaining electrodes are turned off, forming a neck in the column. The reservoir electrode is then activated to pull back the liquid and breaking the neck, leaving a droplet on the electrode predefined. Using this droplet metering structure, droplets down to 20 nl volume can be generated where a deviation may arise, but it is a standard deviation of less than 2 % [5].

A similar technology can be used for the splitting of a droplet into two or more smaller droplets. This may be performed by activating two adjacent electrodes and simultaneously deactivating the electrode holding that droplet before splitting. Since the droplet volume is of great importance for the accuracy of all assays, additional volume control mechanisms are introduced. Two such mechanisms *on-chip capacitance volume control* and *numerical methods for the design of electrowetting-on-dielectric (EWOD) structures* have been proposed [5]. Once the droplets are formed, their actuation is accomplished by the EWOD effect as described above.

The merging of droplets can be achieved easily with the use of three consecutive electrodes. Two droplets are individually moved to electrodes separated from each other by a third one. Deactivating these two electrodes and activating the third separation electrode merging are performed successfully.

Mixing is the most important basic operation in DMFB as proper or improper mixing of samples and reagents may lead to successful or unsuccessful operations. Improper mixing may cause deviation from the result, which may lead to discarding of the chip. The most basic type of mixing within droplets on the EWOD platform is an oscillation, forwards and backwards, between at least two electrodes. Splitting and merging for several times is a type of very efficient mixing that requires three successive electrodes. Another mixing scheme is the repetitive movement of the droplet on a rectangular path. The shortest mixing time for two 1.3-µl droplets in linear oscillation on 4 electrodes is about 4.6 s [5].

2.4 Strengths and Limitations of DMFB

The strengths of the platform are the tiny liquid volumes in the nanoliter range that can be handled with high precision, and the scope to program the droplet movement. This reduces sample and reagent consumption and allows a maximum of flexibility for the implementation of different assay protocols on a chip.

The simple setup without any moving parts can be fabricated using standard lithographic processes. The program-based control of small droplets has its enormous potential, since it allows varying the operations on the same chip. However, although the sample and reagent consumption is low, portable systems, for example, point-of-care applications have not yet been implemented due to the bulky electronic instrumentation required to operate the platform.

Another drawback is the influence of the liquid properties on the droplet transport behavior, that is, different materials show different wetting abilities. This leads to differences in volume or movement speed. Also the long-term stability of the hydrophobic surface coatings is another problem that introduces the contamination risk, since every droplet can contaminate the surface and thus lead to false results. Another issue is the possible electrolysis caused by the electric fields themselves.

In case of handling the fluid, some problems appear. For example, *nucleic acids* are critical molecules because of their negative charge and tendency to adhere to charged surfaces such as metal oxides. Similar problems occur with *proteins* or *peptides* that exist in a variety of electrical charges, molecular sizes, and physical properties. In this case, adsorption is possible onto the surfaces. Again their catalytic (enzymatic) activities can be influenced by the substrate. A general problem due to the interaction of biomolecules and microfluidic substrates is the *blocking* of substrates with another suitable biomolecule that is added in excess.

Instead of the disadvantages, it is highly acceptable and appreciable as the EWOD technique bears great potential to manipulate many single droplets in parallel. Thus, it is of immense importance for today's fast life styles.

3 A Brief Literature Survey on DMFB

At the beginning of this century, the digital microfluidics is being tried to have massive parallelism in bioassay analysis. This parallelism consequently requires concurrent bioassay operations, that is, concurrent movement of multiple droplets throughout a path and/or mixing of two or more reagents and samples in different regions of a bioassay in parallel. Droplets are moved by proper sequence of activation and deactivation of electrodes that are controlled by some external control pins. So, the pins must be so chosen that we can achieve pipelining in droplet routing [12]. In this context, a true parallelism has been introduced in the present article. Now, we survey in very brief how droplets are moved to their destination and tasks are performed accordingly.

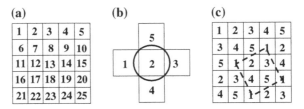

Fig. 2 **a** 25 pins are needed to cover all electrodes of a 5 × 5 array by direct addressing method. **b** Pin number 2 is a droplet holder that has four direct neighbor pins 1, 3, 4, and 5. **c** A 5 × 5 array is covered by five pins using *Connect-5* algorithm

3.1 Direct Addressing Pin Configuration

To move a droplet, activation and deactivation of appropriate electrodes are required [3, 4, 13, 14]. So, every electrode must be controlled by some control pin to provide the necessary actuation voltage. The easiest procedure to assign pins to electrodes is to allot individual control pins. So, the number of pins required for an $n \times n$ array is n^2; a model array is shown in Fig. 2a. A method of partitioning based on array may greatly reduce pin number as stated below.

3.2 Array-Based Partitioning

An array-based partitioning is simple and efficient in respect of the number of distinct voltages we are supposed to provide as input [3, 4, 15–17]. The chip is divided into some partitions depending on the activities performed there, and an optimum number of pins are used to assign the electrodes of the partition. These partitions can be repeated anywhere on a chip to reduce the total number of control pins in the chip. If array-based partitioning is done using *Connect-5* algorithm [4, 13], then we may find that here any pin has four distinct immediate adjacent neighbors; see in Fig. 2b. Thus, we obtain an array of any size by assigning only five pins as shown in Fig. 2c. Though only five pins are sufficient to assign all the electrodes on an array of any size, only a single droplet can safely be allowed to move in such a huge area.

Through the use of *Connect-5* algorithm, electrodes in array of any size can be assigned to pins. Now, if there is more than one droplets to move to different directions, electrode interference may occur. By electrode interference, we mean here that some of the electrodes in the array become activated due to the sharing of a set of five pins by all the electrodes and it results in undesired movement, mixing, splitting of the droplets, or resulting in stuck droplet and thus the performance of the whole chip degrades. In Fig. 3a, there are two droplets each on pin 1 and tends to move to pin 3. As a result, pin 3 is activated simultaneously deactivating pin 1 and both the droplets move to their destined position safely as shown in Fig. 3b.

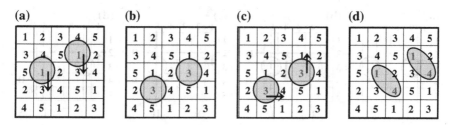

Fig. 3 **a** Both the droplets are on same pin and both of them intend to move to the same pin. **b** Safe movement is possible to pin 3. **c** Both the droplets are on same pin, but tend to move to different directions. **d** Both the droplets stuck between the two diagonally activated electrodes

On the other hand, in Fig. 3c, the droplets are on pin 3 and D_1 is to move rightward on pin 4, whereas D_2 is to move upward on pin 1. So, pins 1 and 4 are activated simultaneously deactivating pin 3. It results in stuck droplets at the junction of pins 1 and 4, as both of them are activated at a time as shown in Fig. 3d. This type of unwanted circumstance is known as electrode interference. As a remedy of this problem, the concept of cross-referencing is introduced.

3.3 Cross-referencing

As a remedy to the problem of using n^2 number of distinct pins for an $n \times n$ array of electrodes, array-based partitioning method is highly efficient [3, 4, 18]. But, electrode constraint is again a hazard to this newly introduced method. Hence, a pin-constrained design technique is introduced namely, cross-referencing [4, 15], where only $m + n$ number of control pins are required to assign to all the electrodes in an $m \times n$ array. In this case, the electrode to be actuated is defined by the row and column number whose intersection contains a next-active (droplet holding) electrode. A next-active electrode is certainly such an adjacent electrode of an electrode that currently holds a droplet.

A method named after *cross-referencing* [4, 15] has been introduced to directly decide the voltage to be applied (HIGH or LOW) at the row and column combination for proper movement of a droplet. Instead of many advantages of this pin assignment technique, there are some disadvantages too. When we activate a row and a column for moving a droplet using HIGH–LOW or LOW–HIGH combination, then some unwanted cells might also be activated that may allow unwanted movement of droplet. The following example of a part of scheduling shows this problem. To authorize only wanted movements, electrode constraints have been introduced accordingly.

HIGH–LOW combinations applied to rows and columns of an array of size 5×5 have been explained in Fig. 4. Say there are two droplets at cells (3,2) and (3,5) as shown in Fig. 4a, and we like to move them to cells (2,2) and (4,5), respectively. A droplet movement is possible only when a LOW–HIGH or a

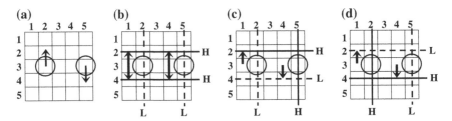

Fig. 4 a The two droplets are moving from their respective cells to the cells pointed by *arrows*. **b** Droplet movement will be in trouble due to electrode interference, as both rows 2 and 4 have been made HIGH. **c** Desired movement as wanted in Fig. 4a is possible if row 2 and column 5 are made HIGH whereas row 4 and column 2 are made LOW. **d** An alternative solution, as desired in Fig. 4a, is obtained when row 4 and column 2 are made HIGH whereas row 2 and column 5 are made low

HIGH–LOW combination is applied for an electrode in a row–column intersection. Hence, if the columns 2 and 5 are made LOW and rows 2 and 4 are made HIGH, then we may find that two desired LOW–HIGH combinations are obtained at cells (2,2) and (4,5) whereas two unwanted combinations are formed at cells (4,2) and (2,5) that confuse the droplets for their desired movements. So, these HIGH–LOW combinations of rows and columns, as shown in Fig. 4b, are not allowable combinations.

On the other hand, there are at least two such desired HIGH–LOW combinations of rows and columns, as shown in Fig. 4c, d, each of which helps in allowing desired movements of the said droplets. So, for the desired movement of the distinct droplets in Fig. 4b, rows and columns may be activated and deactivated as shown either in Fig. 4c or in Fig. 4d to get the desired solution. Incidentally, for a large array with a number of droplets, it has been proved that the problem of satisfying electrode constraints toward a desired solution is an NP-hard problem [3, 15]. Though this is a voltage efficient technique, as a single row and a single column are made either LOW or HIGH, in general for a big array with many droplets, we cannot develop a polynomial time algorithm that expectantly may solve each and every instance of the problem under consideration.

3.4 Broadcasting

In broadcasting, control pins are assigned to electrodes taking into account the movement of droplets that is predefined in terms of scheduling of a complete assay, that is, the activation–deactivation sequence of electrodes [3, 4]. It is stored in a microcontroller in digital term, and the electrodes used to route a droplet is assigned to a control pin maintaining that activation–deactivation sequence. Thus, for a specific bioassay, it reduces the number of pins significantly, and hence, no electrode interference occurs. In case of pin-constrained design, more than one

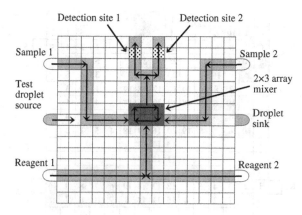

Fig. 5 A 15 × 15 array layout of droplet routing containing two sources of samples (Sample 1 and Sample 2) and two sources of reagents (Reagent 1 and Reagent 2) with one 2 × 3 mixer and two detection sites. Direction of *arrows* shows the movement of droplets along the paths

electrodes are controlled by a single pin. It is voltage efficient, but there is a deficiency that if more than one droplets are to move we have to maintain electrode constraints as well. In this paper, we have adopted the notion of broadcasting to develop a pin configuration of a restricted sized chip for a set of parallel bioassay operations.

4 A 15 × 15 Array and Its Working Principle

4.1 The Existing Bioassay

A DMFB of size 15 × 15 is shown in Fig. 5, where two operations are performed separately on two samples and two reagents [3, 4, 19]. A shared mixer is used, where a first sample (say S_1) and a first reagent (say R_1) are routed from their respective sources to the mixer, and after a desired level of mixing, the mixed droplet is then routed to detection site 1 for necessary finding(s). After completion of this phase, a second sample (say S_2) and a second reagent (say R_2), in a similar manner from their respective origins, route to the mixer for their mixing and then the mixed droplet goes to detection site 2 for necessary outcome(s). So, there should be a delay in between the two operations as the array contains a common mixer, some paths below and above the mixer common to different reagents and mixed droplets to respective detection sites. Now, it is very important that such regions and paths are need to be washed in between every alternative mixing process; otherwise, unwanted contamination of residual samples, reagents, and mixed droplets might cause for erroneous results.

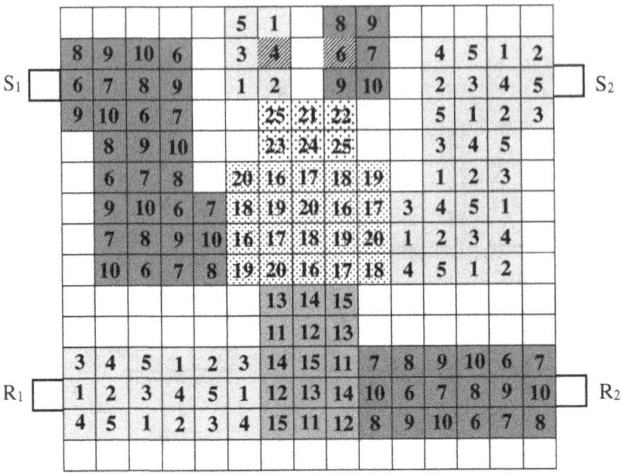

				5	1		8	9						
8	9	10	6	3	4		6	7			4	5	1	2
6	7	8	9	1	2		9	10			2	3	4	5
9	10	6	7			25	21	22			5	1	2	3
	8	9	10			23	24	25			3	4	5	
	6	7	8	20	16	17	18	19			1	2	3	
9	10	6	7	18	19	20	16	17			3	4	5	1
7	8	9	10	16	17	18	19	20			1	2	3	4
10	6	7	8	19	20	16	17	18			4	5	1	2
						13	14	15						
						11	12	13						
3	4	5	1	2	3	14	15	11	7	8	9	10	6	7
1	2	3	4	5	1	12	13	14	10	6	7	8	9	10
4	5	1	2	3	4	15	11	12	8	9	10	6	7	8

(Left labels: S_1 at the third row block, R_1 at the lower block. Right labels: S_2 at the upper block, R_2 at the lower block.)

Fig. 6 Pin assignment of the array using *Connect-5* algorithm that covers all the distinct partitions and uses not more than 25 pins. Here, for the movement of a droplet along a path, the adjacent cells are used as guard cells in most of the regions. Hence, for a mixer of size 2 × 3, an array of size 4 × 5 is deployed for its realization. Different *colors* show the different partitions of the array for pin assignment

Though there are two detection sites, as the mixing is done in sequential order, the chip is underutilized. If we use only one shared mixer, as in the case of existing bioassay, washing of the mixing region as well as the paths below and above the mixer, etc., is to be performed as an intermediate task of the assay. So, in this case, the mixing operation is sequential in nature, where an inter-operational gap for washing is necessary.

To achieve the aforementioned assay operations, the *Connect-5* algorithm is used as pin assignment [3, 4, 13]. Hence, there are at least five independent partitions for desired movement of droplets obeying all necessary tasks as sought by the said assay, and as an optimized result, 25 control pins are required [4, 13, 19], as shown in Fig. 6.

Furthermore, based on a new requirement, say S_2 and R_1 are to be mixed first, and then S_1 and R_2 are to be mixed in another instance of time (from their respective sources, as shown in Fig. 5). Here, also the process is exactly similar (as only one mixer is there in the array) to the earlier process of mixing.

A similar situation occurs when S_1 and R_2 are taken care of for their mixing. Needless to mention that, an immediate washing is necessary between different pairs of consecutive mixing. Thus, using two samples and two reagents, six different combinations of mixing are possible (like S_1–R_1, S_1–R_2, S_2–R_1, S_2–R_2, S_1–R_1–R_2, and S_2–R_1–R_2) for their relevant detections in different instances of time, and all of them can be performed on the given array using appropriate scheduling providing desired stall and necessary washing in between [13, 19].

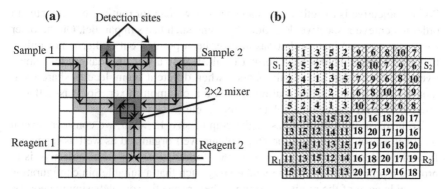

Fig. 7 **a** Layout and droplet routes of the 10×10 array chip with a 2×2 mixer. **b** Partitions along with pin configuration of the 10×10 array that requires 20 pins only. Here, the 2×2 mixer resides at the middle of the array comprising pins 3, 10, 12, and 19, taking one cell from each of the partitions

4.2 A Modification Over the 15 × 15 Array

A modification of the previous array (in Fig. 6) has been proposed by Hwang et al. [2, 20], where the array size is reduced to 10×10, the number of partitions is reduced to four, and the mixer size is 2×2 (instead of 2×3) as shown in Fig. 7a. Though this modification reduces the number of pins required but yet the mixing or detection is sequential in nature as the number of mixers is not increased. In Fig. 7b, we may observe that the mixing region of the array is the junction of four partitions (taking only one cell from each of the partitions), so several droplets (as necessary) can move to this region for mixing. Here, the droplets do not suffer by the limitations of *Connect-5* algorithm; rather, the pin number is also reduced by 20 % (as only four instead of five modules are present in this modification over the earlier array).

4.3 Requirements and Objectives Toward Parallelism

In the earlier work, we have seen that using an array of size 15×15, we acquire only one bioassay outcome at a time, where only 58 cells are used for routing, mixing, and detection, 91 cells are used as guard cells, and the remaining 76 cells are fully unexploited (see Fig. 6). As usual, the bioassay completion time is dominated by the time required for mixing where routing time is negligibly small (1,000:1, or more), as more time devoted for mixing results more truthful detection of some parameter(s) of a sample. In this case, for any pair of combinations of sample and reagent, assay operations are executed one after another.

We know that the *Connect-5* algorithm has its inherent limitations [13]. This is not a voltage efficient design as well as wasteful from the utilization of array space.

So, our objective is to activate as many cells as we can for routing, mixing, etc., in order to achieve respective detections by doing such tasks in parallel. On the other hand, cross-contamination that has already been pointed out earlier is not only a major problem in achieving a correct result, it may also cause danger for human lives. We know that it can only happen when different biomolecules share common cells in their path for routing, or share a common mixer. So, it is better to provide disjoint routes of droplets, if applicable.

To increase the effectiveness of the chip of size 15×15, we consider several practical aspects by making the chip voltage well organized as well as ingenious from the utilization of array space. In this formulation, our focal objective is to provide a proper division of the whole array such that a suitable pin configuration may guide each of the partitions to move their respective droplets simultaneously. Moreover, if necessary, we like to introduce two or more droplets to be inserted into a partition for their synchronized movement as needed. In this paper, we also like to introduce the scheduling of wash droplets by modifying the whole design assay, as and when necessary, to make the outcomes exclusively useful.

5 A New Algorithm for Parallel Bioassay Operations

In this section, we have developed a novel algorithm for accomplishing several pairs of bioassay operations in parallel.

5.1 An Algorithm for Pin Assignment

As *Connect-5* algorithm tolerates unwanted movement of droplets that creates lots of problem in executing the tasks, hence a different pin assignment is a matter of research, which may satisfy some well-defined objectives taking care of all practical hazards that may arise. Of course, minimization of pin number along with its proper assignment over a restricted sized chip, here 15×15, may be considered as a goal.

The chip can be used more resourcefully if we be able to design the chip for mixing separate sets of samples and reagents in isolation in two different mixers and route them to the detection sites at the same time. This is how delay can also be minimized. So, this alteration can give two results in parallel. Let us consider the droplet routing layout as shown in Fig. 8, where two independent mixers (M_1 and M_2) have been introduced. Our objective is to efficiently use the restricted size chip, as well as to assign all the electrodes using a minimum number of electrodes possible. In this context, we focus mainly on the parallelism that can be incorporated in the tasks to be performed.

Module placement: Here, we have two sample and two reagent sources. Again if we complete two mixings simultaneously, we have to perform the detections

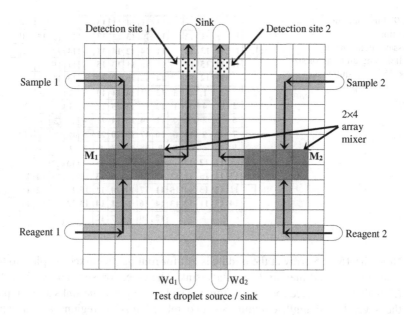

Fig. 8 Layout for droplet routing of a modified 15 × 15 array, where two 2 × 4 mixers are used to execute two mixing operations in parallel. Two sources (and sinks) of test droplet and/or wash droplet are placed at the *bottom* of the array. Droplet sinks are placed on the top of the array where mixed droplets are thrown away from the array after desired detections

individually in two separate detection sites and then they are to be moved to two wastage sinks. To make the chip reusable, we need to perform washing of the used paths and mixers immediately after one use of the chip such that it may be free from cross-contamination. If the washing is also arranged in parallel, two wash droplet sources are to be used. From all the assumptions, we can conclude that we need to place eight devices (sources and sinks) at the periphery of the 15 × 15 chip and four modules (two mixers and two detection sites) are to be placed inside the chip. Again to ensure parallelism as well as to avoid cross-contamination, the routing path of the samples and reagents is to be set as much disjoint as possible. These are put into consideration at the time of placement.

Here, we have planned to place the devices and the modules as shown in Fig. 9, as two parallel assay operations are the target of this design and the whole chip is to be partitioned into two halves. The key ideas are as follows:

- The 15 × 15 array is subdivided into three arrays of 15 × 5. We place the mixing regions in the first and third array starting from the left and one sample and reagent source pair is placed at the left periphery while another pair is placed at the right periphery of the chip. The paths from left (right) sources to the left (right) mixer are almost equal that helps to schedule the mixing concurrently.

Fig. 9 Pin configuration for executing two sets of bioassay operations in parallel using an array of size 15 × 15 that requires no more than 21 pins

S$_1$ (row 3, left) · S$_2$ (row 3, right) · R$_1$ (row 13, left) · R$_2$ (row 13, right)

					13	11	14	11	13					
	14	15	11	12	12	15	13	15	12	12	11	15	14	
	11	12	13	14	11	14	12	14	11	14	13	12	11	
	13	14	15	11	15	13	11	13	15	11	15	14	13	
		11	12	13	14	12	15	12	14	13	12	11		
	5	1	2	3	13	11	14	11	13	3	2	1	5	
2	3	4	5	1	12	15	13	15	12	1	5	4	3	2
5	1	2	3	4	11	14	12	14	11	4	3	2	1	5
7	8	9	10	6	16	17	19	20	21	6	10	9	8	7
9	10	6	7	8	9	18	16	18	9	8	7	6	10	9
		11	12	13	15	20	5	17	15	13	12	11		
13	14	15	11	12	13	21	6	16	13	12	11	15	14	13
11	12	13	14	15	11	17	19	20	11	15	14	13	12	11
14	15	11	12	13	14	16	7	21	14	13	12	11	15	14
					12	20	8	17	12					

- Now, the 15 × 5 array at the middle is left for routing the mixed droplets to the detection site and then to the wastage sink. The detection sites are at (2,7) and (2,9), that is, one electrode apart from the sink if we place the sinks at the top of the seventh and ninth column. So, two related mixing regions are at same distance from them. We have not placed it on the topmost electrode of the corresponding column, as mixed droplets are usually double droplets, which may have more overlapping to the adjacent electrodes. This overlapping to the sink is not desirable.

- Wash droplet sources are placed at the bottom of the columns seven and nine, such that two associated mixing regions are at same distance from them.

The chip can be used more resourcefully if we be able to design the chip for mixing separate sets of samples and reagents in isolation in two different mixers and route them to the detection sites at the same time. This is how delay can also be minimized. So, this modification can give nearly two results in parallel. Let us consider the droplet routing layout as shown in Fig. 8, where two independent mixers (M$_1$ and M$_2$) have been introduced: M$_1$ for mixing S$_1$ with R$_1$, and M$_2$ for mixing S$_2$ with R$_2$. As S$_1$ and R$_1$ are to be mixed in M$_1$, it is desirable if the coupled droplets could be moved simultaneously as much as we can. This is equally true for the sample–reagent pair S$_2$ and R$_2$ that is supposed to be mixed in M$_2$. Hence, for all these droplets, same sequence of pin configurations numbered 1–10 is adopted as we observe in Fig. 9.

As just mentioned that a separate set of pins is not required for the mixing regions, where mixing is performed in both the mixers at the same time instance. So, these pins are also arranged and through activation and deactivation of electrodes in adjacent cells, droplets are moved inside each mixer, as we desire to move them. Out of several possible ways of mixing, a zigzag mode of mixing [21, 22] is shown in Fig. 10.

If there are two reagents and two samples provided, there are 6 possible combinations of mixing as there is no need to mix sample–sample or reagent–reagent pair.

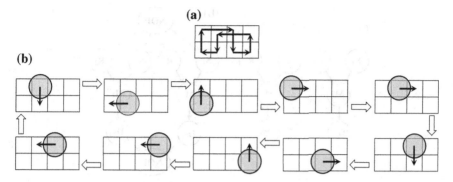

Fig. 10 a The routing path of a mixed droplet in a zigzag mode of mixing inside a 2 × 4 mixer that has been introduced in duplicate for separate mixing of mixed droplets. Here, the *arrows* show the transport pattern of partially mixed droplets for proper mixing. **b** The steps of one complete cycle of mixing showing the movement of a mixed droplet inside a mixer

So, we are interested in performing two mixings concurrently such that throughput is doubled than the previous case where a single mixer has been used. To do it, we are somewhat tricky in assigning the pins to the mixing region. We use same set of pins, which are mirror refection to one another in the two mixers.

Again to minimize the pin count, our objective is to schedule the assay in such a way that the reagent and the sample droplets can reach the mixer within a specified time for mixing and the chip utilization is maximized. The pin configuration has been shown in Fig. 9. So, these pins are also set and through activation and deactivation of electrodes in adjoining cells, droplets are moved within each mixer, as we want to move them. Out of several possible ways of mixing, a zigzag mode of mixing [6, 7] is shown in Fig. 10. It may be noted that a same set of pins, 11–15 applied for routing of S_1, S_2, R_1, and R_2 in synchronism, can also be utilized for the movement of mixed droplets from their mixers to the succeeding detection sites.

We have used the set of pins comprising 11–15 for four 5 × 5 arrays at the four corners of the 15 × 15 array and *Connect-5* algorithm is operated. These arrays are employed only for routing of the droplets to be mixed separately in two mixers maintaining the sequence graph. Pins 1–5 have been assigned to two 3 × 5 arrays along the rows, sixth, seventh, and eighth in the first five columns and in the last five columns, and pins 6–10 have been assigned to two 2 × 5 arrays along the ninth and tenth row in the first five columns and in the last five columns using *Connect-5* algorithm again.

The path of mixed droplets from the mixers to the detection sites is also assigned with the same set of pins (11–15) except at eighth row, where pins from set 1–5 have been used. A cell of this region is activated only when the preferred level of mixing is over and mixed droplets are ready to move to the detection sites. It may further be noted that for mixing in M_1 and M_2, 10 separate pins (two sets of 5 pins) 1–5 and 6–10 are introduced that are washed (by routing wash droplets) in

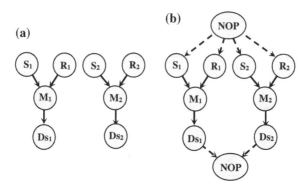

Fig. 11 a The sequencing graph of the assay operations, where S_1 and R_1 (S_2 and R_2) mix in M_1 (M_2) and detection is performed in Ds_1 (Ds_2). **b** An associated sequence graph of the assay operations

between routing of unlike reagents that may use this set of pins as a common portion of their paths. Now, for arranging all types of mixing among two sets of samples and reagents and to reuse the mixing regions as early as possible, washing is performed at the earliest. So, wash droplet paths are assigned with a different set of pins (16–21) such that they have no overlapping with the regular droplets. But, the two wash droplets destined for the two mixers can move concurrently.

This is how we obtain a complete pin configuration for executing all tasks associated in performing some bioassay operation in parallel using a biochip of size 15×15, where at most 21 pins are required, which is not more than the pin number that Chakrabarty et al. used in performing the tasks one at a time for an array of size 15×15 [5, 10]. The above pin configuration is shown in Fig. 9.

5.2 Working of the Chip

Different combinations of accessible mixing make the chip highly favorable. In the following paragraphs, we like to describe in detail the performance of the chip.

We may examine that there are six sets of possible blend of mixing of sample(s) and reagent(s) as follows that could be executed concurrently using the array designed above: $\{(S_1,R_1), (S_2,R_2)\}$, $\{(S_1,R_2), (S_2,R_1)\}$, $\{(S_1,R_1), (S_1,R_2)\}$, $\{(S_2,R_1), (S_2,R_2)\}$, $\{(S_1,R_1), (S_2,R_1)\}$, and $\{(S_1,R_2), (S_2,R_2)\}$.

Let us consider the first case. In this case, the samples and reagents S_1, S_2, R_1, and R_2 can move simultaneously as their paths do not cross (or overlap) from their sources to the relevant mixers. S_1, S_2, R_1, and R_2 are dispensed concurrently, and S_1 and S_2 follow the path 11–12–13–15–12–1–4–2, while R_1 and R_2 traverse the path 11–12–13–15–12–6–8–2 in 8 clock pulses, that is, at clock pulse 8, S_1 and R_1 reach to the electrode assigned by pin 2 at mixer M_1, and S_2 and R_2 reach to the electrode assigned by pin 2 at mixer M_2 and subsequently mixing starts in

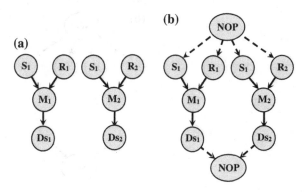

Fig. 12 **a** The sequencing graph of the assay operations, where S_1 and R_1 (S_1 and R_2) mix in M_1 (M_2) and detection is performed in Ds_1 (Ds_2). **b** An associated sequence graph of the assay operations

separation in two different mixers. So, in this case, there is no need to incorporate stall to any of the droplets to be mixed. But, in each of the remaining cases suitable stall is to be provided before mixing of desired droplets at some preferred mixer. Figure 11 shows the corresponding sequence graph for the intended set of operations.

Let us now consider the third case when S_1 is mixed with R_2 in M_2, and also S_1 is mixed with R_1 in M_1. The corresponding sequence graph is shown in Fig. 12. In this case, at first, a droplet of S_1 is routed from its own source to M_2 using the path 11–12–13–15–12–1–4–5–1–12–11–16–17–19–20–21–6–4–3–2 and thus reaches at mixer M_2. With proper stall, R_1 and R_2 are routed from their sources to M_1 and M_2 for mixing in the relevant mixer (at cell 2). Again a second droplet of S_1 is also dispensed at the same time instant with R_1 and R_2, such that mixing in both the mixers may be completed simultaneously. More specifically, in the path of S_1, when the droplet is at cell 6, [in cell location (9,11)], droplets of R_1, R_2, and the second droplet of S_1 are entered in the array activating pin 11 at locations (13,1), (13,15), and (3,1), respectively. From their corresponding positions, R_1 and R_2 traverse the related path 11–12–13–15–12–6–9–2 and S_1 is moved on the path 11–12–13–15–12–1–4–2. Thus, after 8 clock pulses of their entrance to the array, these three droplets reach their own destination and become ready for carrying out a synchronized course of mixing in both the mixers.

So, it is required that the pair of droplets to be mixed in each mixer may arrive the starting point, that is, on pin 2 in the mixer, at a time avoiding all the electrode interferences in the path of its own and that of in the paths of other droplets to reach to their mixers. To satisfy this condition, when R_1, R_2, and the next droplet of S_1 are moved on pin 12, the first droplet of S_1 is also moved to pin 4, that is, upward from pin 2. In this way, when the second droplet is on pin 1, first droplet is also on pin 1 to synchronize the movement and then all the four droplets reach the starting electrode of mixing, that is, on pin 2 simultaneously. At this moment, the harmonized mixing (in both the mixers) starts and then the matched steps of

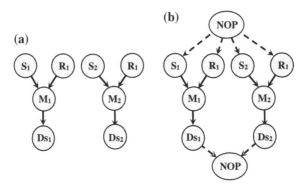

Fig. 13 **a** The sequencing graph of the assay operations, where S_1 and R_1 (S_2 and R_1) mix in M_1 (M_2) and detection is performed in Ds_1 (Ds_2). **b** An associated sequence graph of the assay operations

detection, etc., are performed in parallel, as desired in a case of application. Note that here intermediate washing on the path of droplet S_1 is not required; as in both the mixers, the reagents are to be mixed with the same sample. So, there is no question of contamination. Figure 12 shows the sequence graph required for the assay operation.

Another combination of mixing is S_2–R_1 in M_1 and S_2–R_2 in M_2, that is, S_2 is common in the two mixing. So, S_2 is to be distributed to the mixer instead of S_1 (in the previous case). We can follow a similar schedule where S_2 is entered first and then the other three droplets are entered and moved accordingly as has been stated in the previous case (for S_1).

There is another combination, where two samples are to be mixed with single type of reagent (R_1 or R_2), which has been shown using the sequence graph in Fig. 13. In this case, we distribute the common reagent to both the mixers. Let us consider that R_1 is to be moved to M_1 and M_2. Accordingly, R_1 is dispensed at (13,1) and traversed the path 11–12–13–14–15–11–17–19–20–16–17–18–20–21– 6–10–9 and reaches to M_2. In the mean time, when the droplet is on pin 20 [at location (9,9)], S_1, S_2, and the second droplet of R_1 are entered into the array at locations (3,1), (3,15), and (13,15), respectively, by activating pin 11 and then they are moved to their destined mixer. Their paths are same as earlier. As they reach the mixers at a time, mixing may be performed fully in parallel. Again for the sake of synchronism, when there are two droplets in a partition assigned to pins using *Connect-5* algorithm, it is obvious to have unidirectional movement of the droplets for their safety. In this case also, when in mixer M_1, 6 is activated to hold the second droplet of R_1, the first droplet of R_1 in mixer M_2 is also moved to pin 6, and then both of them move on pin 9 when S_1 and S_2 are on pin 4. At the next clock pulse, the entire four droplets move on pin 2 and thus mixing is started in both the mixers in parallel.

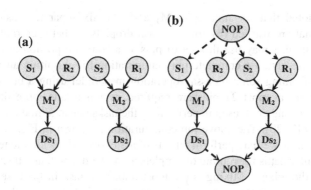

Fig. 14 **a** The sequencing graph of the assay operations, where S_1 and R_2 (S_2 and R_1) mix in M_1 (M_2) and detection is performed in Ds_1 (Ds_2). **b** An associated sequence graph of the assay operations

In the same way as has been told in the last paragraph, another assay may be accomplished on this platform, which requires the mixing of S_1–R_2 in M_1 and S_2–R_2 in M_2.

We may think that the second case of mixing is the most cumbersome, when we mix S_1 with R_2 (in M_1) and S_2 with R_1 (in M_2). Figure 14 shows the corresponding sequence graph (individually for two mixers and the joint sequence graph).

We briefly state it as follows. In this case, a droplet of reagent R_1 is first dispensed from its source, and it follows the path 11–12–13–14–15–11–17–19–20–16–17–18–20–21–6–10 for mixing. When this droplet is at cell 20 (in row 13), then a wash droplet WD_1 is entered into the array at row 15 activating pin 20 and R_2 is ready for its entry into the array. Note that the wash droplet, WD_1 moves through the path 20–16–17–19–20–21–17 to the right source/sink of wash droplet. In this path of WD_1, when it is just out of the array and on the other hand R_1 droplet is on pin 6 [at location (9, 11)], then cell 11 is activated to enter the droplets of S_1 and S_2. At this very moment, S_1 and S_2 are entered into the array from their respective source, and at this point in time R_2 resides at cell 20 at location (13,10) whereas R_1 is at cell 6, and then it is stalled at pin 10 [at cell location (11, 7)]. Thus, R_2 moves to cell 10 of mixer M_1. When R_2 is to be entered in M_1, as R_1 is already waiting in M_2 and both the mixers are assigned to same set of pins, R_1 and R_2 must be synchronized. So, when R_2 appears on pin 6, R_1 is also moved to pin 6 from pin 10 to avoid electrode interference. On the other hand, S_1 and S_2 reach to mixers M_1 and M_2, respectively, and pairwise all the four droplets appear on pin 2 in the two mixers at a time. As a consequence, mixing starts and continues in complete synchronization.

We may further notice that there are decision points (or decision regions) comprising four cells 11, 14, 16, and 17 adjacent to M_1 and 11, 14, 20, and 21 adjacent to M_2 for the movement of mixed droplets from the respective mixers to the detection sites. A cell of this region is activated only when the desired level of mixing is over and mixed droplets are ready to move to the detection sites. It may further be noted that for mixing in M_1 and M_2, six separate pins 16–21 are introduced that are washed (by routing wash droplets) in between routing of different reagents that may use this set of pins as a common portion of their paths. This is how we obtain a complete pin configuration for executing all tasks associated in performing some bioassay operation in parallel using a biochip of size 15×15, where at most 21 pins are required, which is one more than the pin number that Hwang et al. used in performing the tasks one at a time for an array of size 10×10 [2, 20]. The above pin configuration is shown in Fig. 9.

So, this is how we may perform all sets of assay operations mentioned above in parallel. Similar paths for routing of droplets are to be defined, and then, mixing is executed in the mixers. Mixing is the most important task in the case of droplet routing and the most time consuming step in the whole process. Here, we use mixers of size 2×4 each and allow zigzag way of doing the task. Then, mixed droplets are sent to the detection sites for detection and thrown away from the array chip as a later step.

5.3 An Example Run of Assay Operations in Parallel

As an example run, here we consider a typical case of parallel mixing as shown in Fig. 15, where R_1 mixes with R_2 and S_2 mixes with R_3; as already discussed several cases above, here we place the source of R_3 instead of R_2 for both the phases and also we place R_2 instead of the source of S_1 only for the first phase of mixing, and then the source of S_1 is placed as usual for the second phase. One part of mixed droplet of R_1 and R_2 mixes with S_1; on the other hand, another part of mixed droplet of R_1 and R_2 mixes with one part mixed droplet of S_2 and R_3. While one part of mixed droplet of S_2 and R_3 mixes with a part of mixed droplet of R_1 and R_2, at that time another part of mixed droplet of S_2 and R_3 reaches to the detection site (Ds_2) for detection. Subsequent to the second phase of mixing, the mixed droplet of M_1 and M_2 reaches to the detection sites Ds_1 and Ds_2, respectively, for detection. The algorithm has already been made earlier, and thus, we obtain the scheduling map, which is the necessary final outcome of the assay operations performed in a given array of size 15×15.

We know that every operation is having its own sequence of processing. As for example, mixing of S_2 and R_3 in M_2, and then subsequent detection of the mixer in the detection site (Ds_2) is a complete course of action; an allied graphical representation of this fact is shown in Fig. 15a, which is an example of a sequencing graph.

A sequencing graph may contain two or more connected components as distinct assay operations that are usually disjoint from each other. If we like to perform some of the sequencing steps in parallel, then level by level that information is captured when we represent it with the help of a connected sequence graph. As for example, as shown in Fig. 15b, the graph tells that the droplets of samples and

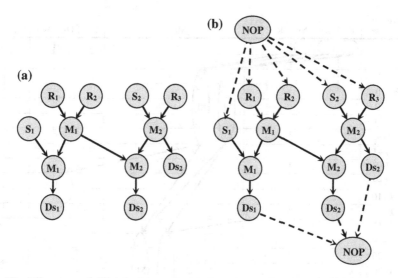

Fig. 15 **a** The sequencing graph of the assay operations, where R_1 and R_2 (S_2 and R_3) mix in M_1 (M_2) and detection is performed in a later phase (Ds_2). Then, in a next phase, S_1 is mixed with the mixed droplet of R_1 and R_2 (the mixed droplet of S_2 and R_3 is mixed with the mixed droplet of R_1 and R_2) in M_1 (M_2) and detection is performed in Ds_1 (Ds_2). **b** An associated sequence graph of the assay operations

reagents are dispensed and routed altogether, mixed concurrently at their respective mixers, and also detected at the same time in the coupled detection sites.

Here, we may note that there is no point of cross-contamination as the paths for samples and reagents are distinct, and there is no overlapping even for a single cell. So, washing is not required; however, in between two sets of consecutive assay operations, it is desirable to wash the whole path and also the mixers, detection sites, etc. The scheduling map for these assays is shown in Fig. 16.

We also may observe that here in this structure of pin assignment, as we execute assay operations in parallel, we gain $\sim 100\,\%$ time (or more) in detecting the samples. Use of wash droplet(s) might require in performing other pairs of assay operations where a cell or a series of consecutive cells in a path is required to wash within the assay operations in parallel. In these assay operations, we may observe that, 83 cells are used for routing, mixing, and detection, 96 cells are used as guard cells, and the remaining 46 cells are totally inoperative (see Fig. 6).

5.4 Experimental Results

In this section, we compare the three biochips, two in existing articles [2–4, 13, 19, 20] and the one introduced in this paper, as shown in Figs. 5, 6, 7, 8, and 9, from their structural and functional points of view. The primary differences are

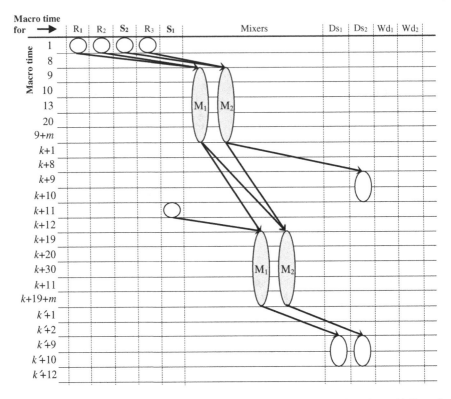

Fig. 16 The scheduling map for the sequencing graph in Fig. 15a, where R_1 mixes with R_2 and S_2 mixes with R_3, and then one part of the mixed droplet of R_1–R_2 is mixed with S_1 in M_1 and the other part of the mixed droplet of R_1–R_2 is mixed with the mixed droplet of S_2–R_3. Here, m is the number of clock pulses required for mixing, and $k = 9 + m$ and $k' = k + 19 + m$. Moreover, at the $(k + 12)$th and $(k'+12)$th clock pulse, the mixed droplets are thrown away from the array after the detection is made at the $(k + 9)$th and $(k'+9)$th clock pulse, respectively

whether two bioassay operations are performed sequentially or in parallel, what are the sizes of mixers, amount of mixing time needed, utilization of cells in an array, number of tasks carried out, etc. These are all tried to include in Table 1 and thus explained in brief as follows.

Regarding the number of cells in a biochip, 33.78 % cells are unused in the foremost biochip whereas this value is only 20.44 % in our design. In our biochip, we introduce 16 % less pins though its achievement is more than 100 % that uses two larger mixers. This happens in practice because the former biochip needs 12 clock pulses for routing of sample and reagent droplets from their sources to the mixer and 19 clock pulses for routing of mixed droplet from mixer to the detection site and disposing, making a total of $2 \times (31 + m + d) + w$ clock pulses for a pair of two successive assay operations with inter-assay washing, whereas this value reduces to only $15 + m + d + (w)$ in our configuration, where no additional clock pulses are applied for washing the biochip during the bioassays are executed in parallel.

Table 1 A table of comparison that assesses two existing arrays and the array introduced in this paper from their pattern and practical viewpoint

Array structure	15 × 15 (Fig. 6)	10 × 10 (Fig. 7)	15 × 15 (Fig. 9)
Mode of operation	Sequential	Sequential	Parallel
# of tasks	Six	Five	Eight
# of mixers	One	One	Two
Mixer size	2 × 3	2 × 2	2 × 4
Pin count	25	20	21
# of active cells	58	48	83
# of guard cells	91	46	96
# of unused cells	76	6	46
Wash droplets	No	No	Yes
# of clock pulses (for two assays)	$2 \times (12 + 19 + m + d) + w$	$2 \times (8 + 5 + m + d) + w$	$7 + 8 + m + d + (w)$

Here, m, d, w, and (w) are the number of clock pulses applied for mixing, detection, inter-assay washing, and intra-assay washing, respectively

In both the earlier two biochips, the primary tasks carried out are routing, mixing, and detection, where sample and reagent droplets are dispensed and routed up to mixer, mixed droplets are routed from the mixer to the detection sites, and detected droplets are (routed and then) thrown away from the biochip. In our bioassay, we have included two more tasks: One is inclusion of decision point (in a decision region) after each mixer, a cell of which is activated only when the mixing operation is over, and the second one is the routing of wash droplets as intra-assay requirement, which is mandatory but not mentioned in both the earlier works.

6 Conclusion

In this paper, a restricted sized biochip whose array size is 15 × 15 has been taken under consideration. In existing literature, such an array is used for only one bioassay operation at a time as there is only one mixer of size 2 × 3. This chip is underutilized that had been pointed later and subsequently a 10 × 10 array is introduced with a mixer of size 2 × 2, though the assay operation is still sequential as they are sharing a single mixer. In all these respects, we have configured a pin assignment where the pin count is 16 % less than that of the earlier chip of same size, but here the assay operations are performed in parallel where washing as cross-contamination avoidance has also been introduced. In such a chip, we have introduced an additional region as decision point, which is activated only when a desired level of mixing is over. We strongly guess that intensive research could be carried over in future on decision point as an important task in between mixing and detection in any assay operation. In our configuration, larger mixers have also been included wherein we suggest zigzag way of mixing of droplets, which is certainly better even if the same number of clock pulses is applied as the scope of diffusion between the samples and reagents to be mixed is higher in this case. Out of many other achievements, our designed chip configuration achieves more than 100 % outcomes using the same number of clock pulses as the assay operations are executed in parallel.

References

1. http://www.tutorgig.com/encyclopedia
2. Advanced Liquid Logic. http://www.liquid-logic.com
3. Chakrabarty, K., Su, F.: Digital Microfluidic Biochips: Synthesis, Testing, and Reconfiguration Techniques. CRC Press, Boca Raton (2007)
4. Chakrabarty, K., Xu, T.: Digital Microfluidic Biochips Design Automation and Optimization. CRC Press, Boca Raton (2010)
5. Fair, R.B.: Digital microfluidics: is a true lab-on-a-chip possible? In: Microfluid Nanofluid, vol. 3, pp. 245–281. Springer, Berlin (2007)
6. Zeng, J., Korsmeyer, T.: Principles of droplet electro-hydrodynamics for lab-on-a-chip. Lab Chip **4**, 265–277 (2004)
7. Fair, R.B., Srinivasan, V., Ren, H., Paik, P., Pamula, V., Pollack, M.G.: Electrowetting based on chip sample processing for integrated microfluidics. In: IEDM, pp. 779–782 (2003)
8. Srinivasan, V., Pamula, V.K., Pollack, M.G., Fair, R.B.: A digital microfluidic biosensor for multianalyte detection. In: Proceedings of IEEE MEMS Conference, pp. 327–330 (2003)
9. Paik, P., Pamula, V.K., Fair, R.B.: Rapid droplet mixers for digital microfluidic systems. Lab Chip **4**, 253–259 (2003)
10. Su, F., Hwang, W., Chakrabarty, K.: Droplet routing in the synthesis of digital microfluidic biochips. In: DATE, pp. 323–328 (2006)
11. Böhringer, K.F.: Towards optimal strategies for moving droplets in digital microfluidic systems. In: ICRA, pp. 1468–1474 (2004)
12. Chakrabarty, K.: Digital microfluidic biochips: a vision for functional diversity and more than Moore. In: VLSI Design, pp. 452–457 (2010)
13. Xu, T., Chakrabarty, K.: Automated design of digital microfluidic lab-on-chip under pin-count constraints. In: ISPD, pp. 190–198 (2008)
14. Zhao, Y., Chakrabarty, K.: Pin-count-aware online testing of digital microfluidic biochips. In: IEEE VLSI Test Symposium, pp. 111–116 (2010)
15. Xu, T., Chakrabarty, K.: A droplet-manipulation method for archiving high throughput in cross-referencing based digital microfluidic biochips. TCAD **27**, 1905–1917 (2008)
16. Xu, T., Chakrabarty, K.: Droplet-trace-based array partitioning and a pin assignment algorithm for the automated design of digital microfluidic biochips. In: IEEE/ACM ICH/SCSS, pp. 112–117 (2006)
17. Xu, T., Hwang, W.L., Su, F., Chakrabarty, K.: Automated design of pin-constrained digital microfluidic biochips under droplet-interference constraints. ACM J. Emerg. Technol. Comput. Syst. **3**(3), Article 14 (2007)
18. Xu, T., Chakrabarty, K.: A cross-referencing-based droplet manipulation method for high-throughput and pin-constrained digital microfluidic arrays. In: DATE, pp. 552–557 (2007)
19. Su, F., Chakrabarty, K.: High-level synthesis of digital microfluidic biochips. In: ICCAD, vol. 3, no. 4, Article 16 (2008)
20. Hwang, W.L., Su, F., Chakrabarty, K.: Automated design of pin-constrained digital microfluidic arrays for lab-on-a-chip applications. In: DAC, pp. 925–930 (2006)
21. Paik, P., Pamula, V.K., Pollack, M.G., Fair, R.B.: Electrowetting based droplet mixers for microfluidic systems. Lab Chip **3**, 28–33 (2003)
22. Paik, P., Pamula, V.K., Fair, R.B.: Rapid droplet mixers for digital microfluidic systems. Lab Chip **3**, 253–259 (2003)

A 2D Guard Zone Computation Algorithm for Reassignment of Subcircuits to Minimize the Overall Chip Area

Ranjan Mehera, Arpan Chakrabarty, Piyali Datta and Rajat Kumar Pal

Abstract The guard zone computation problem finds vast applications in the field of VLSI physical design automation and design of embedded systems, where one of the major purposes is to find an optimized way to place a set of 2D blocks on a chip floor. In VLSI layout design, the circuit components (or the functional units/ modules or groups/blocks of different subcircuits) are not supposed to be placed much closer to each other in order to avoid electrical (parasitic) effects among them (http://en.wikipedia.org/wiki/Curve_orientation, [13]). The (group of) circuit components on a chip floor may be viewed as a set of polygonal regions on a two-dimensional plane. Each (group of) circuit component(s) C_i is associated with a parameter δ_i such that a minimum clearance zone of width δ_i is to be maintained around C_i. The regions representing the (groups of) circuit components are in general isothetic polygons, but may not always be limited to convex ones. The location of the guard zone (of specified width) for a simple polygon is a very important problem for resizing the (group of) circuit components. In this paper, we have developed an algorithm to compute the guard zone of a simple polygon as well as to exclude the overlapped regions among the guard zones, if any. If the number of vertices in the given polygon is n, then our algorithm requires $O(n \log n + I \log n)$ time, where I is the number of intersections among the guard zones. So, it is output sensitive in nature that depends on the value of δ_i. The algorithm developed in the paper is proved to report a preferred guard zone of the given simple polygon excluding all the intersections, if any.

R. Mehera (✉) · A. Chakrabarty · P. Datta · R.K. Pal
Department of Computer Science and Engineering, University of Calcutta,
92, A.P.C. Road, Kolkata 700 009, West Bengal, India
e-mail: ranjan.mehera@gmail.com

A. Chakrabarty
e-mail: arpan250506@gmail.com

P. Datta
e-mail: piyalidatta150888@gmail.com

R.K. Pal
e-mail: pal.rajatk@gmail.com

© Springer India 2015
R. Chaki et al. (eds.), *Applied Computation and Security Systems*, Advances in Intelligent Systems and Computing 305, DOI 10.1007/978-81-322-1988-0_11

Keywords Simple polygon · Guard zone · Notch · Placement problem · Circuit component · Convex hull · Concave vertex · Convex vertex

1 Introduction

Guard zone computation problem is well defined in the literature as an application of computational geometry. Often, this problem is also known as safety zone problem [6]. Given a simple polygon P, its guard zone G (of width r) is a closed region consisting of straight line segments and circular arcs (of radius r) bounding the polygon P such that there exists no pair of points p (on the boundary of P) and q (on the boundary of G) having their Euclidean distance $d(p, q)$ less than r.

In case of VLSI layout design as well as in embedded system, a chip may contain several million transistors. To handle this large number of components, the concept of partitioning is introduced that results in a set of blocks along with interconnections among them. The next step is the placement of these blocks of different dimensions. The goal of placement is to find a minimum area arrangement for the blocks that helps to complete interconnections among them. A good routing and circuit performance heavily depend on a good placement algorithm. Placement of modules is an NP-complete problem [5].

The circuit components on a chip floor may be viewed as a set of polygonal regions on a two-dimensional plane. Each circuit component P_i is associated with a parameter p such that a minimum clearance zone of width p must be maintained around that circuit component. The regions representing the circuit components are general polygons that may not always be convex ones.

The location of the safety zone of specified width for a simple polygon is a very important problem for resizing the circuit components. If more than one polygonal regions are close enough, their safety zones overlap, violating the minimum separation constraint among them. Again inside a notch of such a polygonal boundary, if a wide space is available which may accommodate some circuit component, we cannot use that location, as space for routing the connecting wires from those circuit components to the other circuit components, which are placed outside the notch, is not available. Thus, with respect to the problems of resizing in VLSI circuit components, this is the motivation of defining and computing the safety zone of a (simple) polygon [6].

Without loss of generality, we may assume that the geometric shape of a subcircuit is of a simple polygon P, as shown in Fig. 1a, as the design of each subcircuit is already over; otherwise, sometimes, that could also be considered as a part of floor plan to achieve an alternative design. Several algorithms exist to address the area minimization problem in some phase of physical design either in developing an embedded system or planning for a VLSI circuit [5]. In this paper, we have addressed the problem by computing guard zone of each of the subcircuits (rather, for only a simple polygon) and using guard zone computation as a tool

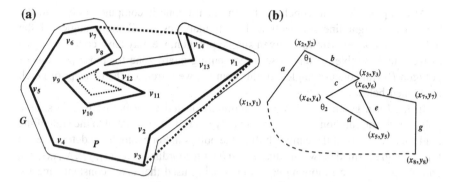

Fig. 1 **a** Guard zone of a polygon. **b** Part of a polygon with vertices (x_1, y_1) through (x_8, y_8), and edges *a* through *g*; the *dotted line* indicates the inner portion of the polygon

toward achieving the same. In Fig. 1a, a 2D simple polygon P is shown by thicker line segments, whose guard zone G is required to be computed outside the polygon.

In this paper, we have proposed and devised an algorithm for computing guard zone of two-dimensional (2D) simple polygon, where each polygon represents a subcircuit such that several such subcircuits are there for their placement on a minimum area chip floor, placing each pair of adjacent subcircuits in a safe separation, toward realizing a VLSI circuit or designing an embedded system. This is necessary to achieve the desired performance of the overall circuit to be designed. In general, an instance of such a problem may contain thousands of subcircuits at some level of design, and the (placement) problem under consideration is NP-complete [5].

There are several other areas, like Robotics, Geographical Information System (GIS), etc., where guard zone computation problem finds its applications.

A simple polygon may contain both convex and concave vertices in it. We define these vertices as follows: A vertex v of a polygon P is defined as convex (concave), if the angle between its associated edges inside the polygon, i.e., the internal angle at vertex v, is less than or equal to (greater than) $180°$. In Fig. 1b, angle θ_1 (between edges *a* and *b*) at vertex (x_2, y_2) is convex, whereas angle θ_2 (between edges *c* and *d*) at vertex (x_4, y_4) is concave.

Let us consider a simple polygon P with n vertices and n edges. A polygon P is given implies that the coordinates of the successive vertices of the polygon are given, where no two polygonal edges cross each other; rather, two consecutive polygonal edges intersect only at a polygonal vertex. We may assume that a portion of this polygon is as shown in Fig. 1b. To know whether an angle θ, inside the polygon, is either convex or concave at vertex v, we do a constant time computation of determining the value (of θ) at vertex v. Thus, all the n internal angles of P are identified as convex or concave in $O(n)$ time.

At this point, we can conclude that the guard zone is computed only with the help of n straight line segments and n circular arcs, if all the n angles of the polygon are convex. But for a given simple polygon, we may have concave angles as well, in P. Problems may arise in computing guard zone for those portions of polygon P with concave angles. In this context, we introduce the concept of notch as defined below.

In R^2, for a given set of three or more connected vertices that form a simple polygon, the orientation of the resulting polygon is directly related to the sign of the angle at any vertex of the convex hull of the polygon. For example, to determine the type of angle formed between edges a and b with coordinates X_A (x_1, y_1), X_B (x_2, y_2), and X_C (x_3, y_3), the following equation is being used that takes constant time for finding out the angle whether it is convex or concave [9].

$$\det(O) = (X_B - X_A)(Y_C - Y_A) - (X_C - X_A)(Y_B - Y_A)$$

The sign of $\det(O)$ helps to identify the type of angle being formed at polygonal vertex X_B. A positive value indicates a convex angle outside the polygon, whereas a negative value indicates a concave angle outside the polygon, and a value zero indicates that the points X_A, X_B, and X_C are colinear. In this way, all the n polygonal angles of P are identified as convex or concave in $O(n)$ time and avoided the use of comparatively costly trigonometric functions.

At this point, it is fair to conclude that the guard zone is computed only with the help of n straight line segments and n circular arcs, if all the n polygonal vertices form convex (external) angles. But for a given simple polygon, it may have concave angles as well, in P. Problems may arise in computing guard zone for these portions of polygon P with concave (external) angles. In this context, the concept of notch has been introduced as defined below.

A *notch* is a polygonal region outside polygon P that is formed with a chain of edges of P initiating and terminating at two vertices of a false hull edge [4]. A convex hull is a convex polygon (having no concave angle) of minimum area with all the points residing on the boundary or inside the polygon for a given set of arbitrary points on a plane.

Clearly, if P is a given simple polygon and CH(P) denotes the convex hull of polygon P, then the area CH(P)—P consists of a number of disjoint notches outside polygon P. According to this definition, a notch is formed outside the polygon in Fig. 2, below (or inside) the dotted line $v_2 v_8$, as this edge is a false hull edge.

Now, the difficulty arises while excluding the part(s) of G that overlap(s). In doing so, we may take the help of digital geometry that can do the task in linear time too [10–12]. But as our inclination in doing the task is by means of computational geometry only, we like to exclude the part(s) of G that overlap(s) using the concept of analytical and coordinate geometry. In the prior computation, we have at most $O(n)$ straight line segments and $O(n)$ circular arcs in computing G. So, to find out all the intersection points and exclude the overlapped region(s) accordingly (in order to obtain only the desired guard zone, including holes, if any,

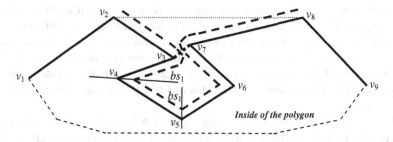

Fig. 2 A notch is formed inside (or *below*) the false hull edge produced by vertices v_2 and v_8, and a guard zone is obtained for this notch as shown by *dotted lines* and *circular arcs* outside of the polygon

as parts of G), we may execute an $O(n^2)$ algorithm for each pair of such segments, among all straight line segments and circular arcs. Indeed, this algorithm is greatly expensive. Hence, while computing the initial guard zone G, we enclose P by G, which is essentially a collection of $O(n)$ line segments only (and there is no circular arc part as we have drawn so far for each of the convex polygonal vertex of P); we explain it in the subsequent sections.

In this paper, we assume that for computing guard zone, an arbitrarily shaped simple polygon is given. A simple polygon is defined as the polygon in which no two boundary edges cross each other. If two or more polygons are close enough so that their guard zones overlap, indicating the violation of the minimum separation constraint among them, the intersecting regions are to be detected such that the guard zone can be computed by eliminating those regions of intersection. In this paper, we have developed an algorithm to compute all these intersections and in the end, the outcome is the targeted guard zone that we compute by eliminating the intersecting regions.

2 Literature Survey

If P is a simple polygon and G is its guard zone of width r, then the boundary of G is composed of straight line segments and circular arcs of radius r, where each straight line segment is parallel to an edge of the polygon at a distance r apart from that edge, and each circular arc of radius r is centered at a (convex) vertex of the polygon. The boundary of a guard zone describes a simple region in the sense that no two edges (straight line segment(s) and/or circular arc(s)) on its boundary intersect in (or pass through) their interior. This has been explained in Fig. 1. The problem originates in the context of resizing of VLSI layout design [7], as described later on (see Sect. 4).

In the context of guard zone computation, several different algorithms have been proposed so far. The most discussed tool for guard zone computation is the Minkowski sum. Apart from Minkowski sum, convolution can also be used as a tool for guard zone computation. A linear time algorithm is developed for finding the boundary of the minimum area guard zone of an arbitrarily shaped simple polygon in [3]. This method uses the idea of Chazelle's linear time triangulation algorithm [2].

Essentially, Minkowski sum between a line (as polygonal segment) and a point (perpendicularly at a distance r apart) with same x- and y-coordinates gives a line parallel to the given one. But the question arises is whether the parallel line is inside or outside the polygon. Here, the definition of Minkowski sum [4] can be extended as below.

If A and B are subsets of R^n, and $\lambda \in R$, then $A + B = \{x + y \mid x \in A, y \in B\}$, $A - B = \{x - y \mid x \in A, y \in B\}$, and $\lambda A = \{\lambda x \mid x \in A\}$. Note that $A + A$ does not equal $2A$, and $A - A$ does not equal 'zero' in any sense.

The convolution between a polygon and a circle of radius r gives us the desired solution. But the circles need to be drawn in every possible point of the polygon, and consequently, the time complexity of the algorithm increases. Minkowski sum and convolution theories find their vast applications in Mathematics, computational geometry, resizing of VLSI circuit components, and in many other subjects/problems.

The computational complexity of the Minkowski sum of two arbitrary simple polygons P and Q is $O(m^2 n^2)$ [1], where m and n are the number of vertices of these two polygons, respectively. In particular, if one of the two polygons is convex, the complexity of Minkowski sum reduces to $O(mn)$. In [3], a number of results are proposed on the Minkowski sum problem when one of the polygons is monotone.

An algorithm for finding the outer face of the Minkowski sum of two simple polygons is presented in [6]. It uses the concept of convolution, and the running time of the algorithm is $O((k + (m + n)\sqrt{l}) \log^2(m + n))$, where m and n are the number of vertices of two polygons and k and l represent the size of the convolution and the number of cycles in the convolution, respectively. In the worst case, k may be $O(mn)$. If one of the polygons is convex, the algorithm runs in $O(k \log^2(m + n))$ time, and there exists no algorithm that can compute the boundary defined by the Minkowski sum of an arbitrary simple polygon and a circle or a convex polygon in time, linear in the worst case of the problem.

In this context, a linear time algorithm is developed for finding the boundary of the minimum area guard zone of an arbitrarily shaped simple polygon in [4]. This algorithm uses the idea of Chazelle's linear time triangulation algorithm [2] and requires space complexity of $O(n)$ as well, where n is the number of vertices of the polygon. After having the triangulation step, this algorithm uses only dynamic linear and binary tree data structures.

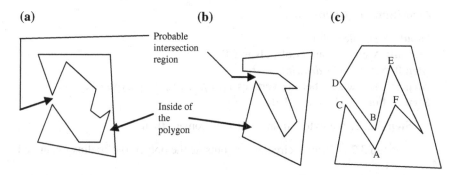

Fig. 3 Different kinds of intersections among guard zones. **a** Overlapping of guard zones that may occur due to two nearby convex regions of a simple polygon. **b** Overlapping of guard zones that may occur due to one *straight line* segment and one convex region of a simple polygon. **c** A simple polygon consists of a number of convex and concave vertices whose guard zonal regions are probable to overlap

3 Formulation of the Problem and the Algorithm

Here, we assume the case of guard zone for a simple polygon. The case becomes simpler if the polygon is convex, and there is no overlapping among the guard zonal regions. Thus, the guard zone for a convex polygon can always be computed in linear time [12]. But for some other cases, a simple polygon may contain notch(s). This is also possible where guard zones for different polygonal segments overlap. Figure 3a, b shows two different cases of such overlapping.

It may so happen that overlapping of guard zones occurs for two close convex polygonal vertices of a simple polygon as shown in Fig. 3a. Again, an overlapping of guard zones may occur for a polygonal edge and a convex polygonal vertex as shown in Fig. 3b.

We can assume another case where a part of a polygon is shown Fig. 3c, where B is a convex vertex and A is concave, and the guard zones of these two regions may overlap. Again the guard zones for the polygonal edges BD and BE may overlap to that of AC and AF.

In this paper, we have developed a sequential algorithm that computes a guard zone G of a simple polygon P. Usually, a guard zone contains straight line segments that are parallel to the edges of P and circular arc-shaped portions of G that are obtained only for the convex polygonal vertices of P. Here, we introduce a variable *optimal-drawing* that we initially set to *true*. During execution of the algorithm, if a concave external polygonal vertex is encountered, the value of *optimal-drawing* becomes *false*. Based on the value of this variable, the algorithm decides whether the intersection step needs to be executed. A *true* value of *optimal-drawing* indicates that the intersection detection step(s) can safely be ignored. Now, the formal steps of the algorithm are as follows:

Algorithm *2D_Guard_Zone*

Input: A simple polygon *P*.
Output: A guard zone *G* of polygon *P*.
Step 0: Set *optimal-drawing = true*
Step 1: Clockwise label the vertices $v_1, v_2, \ldots, v_n$, of polygon *P*.
Step 3: For $i = 1$ to $n - 1$ **do**

> **Step 3.2**: **If** the external angle at v_i is convex **then**
>
> > **Step 3.2.1**: Draw a circular arc (outside the polygon) of radius *r* centered at v_i.
> > **Step 3.2.2**: Find the external angle at v_{i+1}, and consider polygonal edge (v_i, v_{i+1}).
> > **Step 3.2.3**: **If** the external angle at v_{i+1} is convex **then**
> >
> > > **Step 3.2.3.1**: Draw a circular arc (outside polygon) of radius *r* centered at v_{i+1}.
> > > **Step 3.2.3.2**: Draw a line parallel to (v_i, v_{i+1}) at a distance *r* apart from the polygonal edge (outside the polygon) that is a simple common tangent to both the arcs drawn at v_i and v_{i+1}.
> >
> > **Step 3.2.4**: **Else** bisect the external angle at v_{i+1}, denote the bisection bs_{i+1}.
> >
> > > **Step 3.2.4.1**: Draw a line parallel to (v_i, v_{i+1}) at a distance *r* apart from the polygonal edge (outside the polygon) that is a tangent to the arc drawn at v_i and intersects bs_{i+1} at a point, say p_{i+1}.
> > > **Step 3.2.4.2**: Set *optimal-drawing = false*
> >
> > **Step 3.2.5**: Assign $i \leftarrow i + 1$.
> > **Step 3.2.6**: **If** $v_i = v_n$, **then** $v_{i+1} = v_1$.
> > **Step 3.2.7**: **Else** bisect the external angle at v_i, denote the bisection bs_i.
> > **Step 3.2.8**: Find the external angle at vertex v_{i+1}, and consider polygonal edge (v_i, v_{i+1}).
> > **Step 3.2.9**: **If** the external angle at v_{i+1} is convex **then**
> >
> > > **Step 3.2.9.1**: Draw a circular arc (outside the polygon) of radius *r* centered at v_{i+1}.
> > > **Step 3.2.9.2**: Draw a line parallel to (v_i, v_{i+1}) at a distance *r* apart from the polygonal edge (outside the polygon) that intersects bs_i at a point, say p_i, and is a tangent to the arc drawn at v_{i+1}.
> >
> > **Step 3.2.10**: **Else** bisect the external angle at v_{i+1}, denote the bisection bs_{i+1}.
> >
> > > **Step 3.2.10.1**: Draw a line parallel to (v_i, v_{i+1}) at a distance *r* apart from the polygonal edge (outside the polygon) that intersects bs_i at a point, say p_i, and also intersects bs_{i+1} at a point, say p_{i+1}.
> > > **Step 3.2.10.2**: Set *optimal-drawing = false*.

Step 3.2.11: Assign $v_i \leftarrow v_{i+1}$.
Step 3.2.12: If $v_i = v_n$, then $v_{i+1} = v_1$.

End for
Step 4: **If** *optimal-drawing* $=$ *false*, then there is a possibility that two line segments or a line segment and a circular arc or two circular arcs of the guard zone intersect, **then** the line sweep algorithm is executed to determine and report all such intersection points, which eventually exclude the portions of the line segment(s) and/or the circular arc(s) that are at a distance less than r apart from a polygonal edge or a polygonal vertex (outside the polygon).

In this computation of G, we have at most $O(n)$ straight line segments and $O(n)$ circular arcs, if P has no false hull edge. Otherwise, to compute all the intersection points on G and exclude the overlapped region(s) accordingly (in order to get the desired guard zone only, including hole(s), if any, as part(s) of G), we may execute an $O(n^2)$ algorithm for each pair of such segments, among all straight line segments and circular arcs. Indeed, this algorithm is greatly expensive. In this paper, we have developed an algorithm to find the guard zone of a simple polygon efficiently such that all the overlapped regions are excluded, and we have done it using the concept of computational geometry.

A polygon consists of a number of convex and concave vertices. At the convex vertices, the guard zone is circular in shape having a predefined distance r from the polygon vertex [12]. So the circular arc has two tangents at two points where it meets with the two line segments of the guard zone as shown in Fig. 5. In case of guard zone for concave vertices or line segments, we can compute the intersections of the guard zones (if any) by using line sweep algorithm [13] as those guard zonal regions are straight line segments. But in case of circular arc, it cannot be applied so easily as for line sweep algorithm, it is obvious to use starting and end points of the line segments as event points [13]. But a guard zone of a polygon is not only a set of line segments only, there are circular arcs as well. So, line sweep algorithm cannot be applied to it directly.

We know that a circular arc can be assumed as the aggregation of a number of infinitesimally small ordered line segments. Thus, if we can somehow subdivide the arc into line segments, we can then apply the line sweep algorithm for detecting the intersection points between those circular arcs with others. Though to break an arc into line segments, we have to take some precision, that does not affect the output of our algorithm as it definitely reports all the intersections existing among the guard zonal regions.

In our algorithm, we have developed a procedure to subdivide the circular arc into a collection of small arcs such that they can be considered as line segments. If we take n number of iterations to subdivide the arc, we get $(n + 1)$ number of line segments, i.e., $(n + 2)$ number of event points (starting and end points) as two consecutive line segments share one point as starting of one and end point of another. So, $(n + 1)$ line segments have $(n + 2)$ event points. This arc is also to be

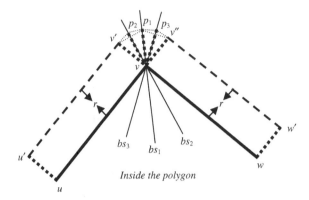

Fig. 4 Recursive division of a convex polygonal angle formed at vertex v wherefrom equal-length smallest possible chords are computed that in a group replaces the circular arc that is computed as a part of guard zone outside the polygon up to a desired level of precision of an angle that is formed at the convex vertex for each such smallest possible chord

checked with other guard zonal arcs and line segments for intersections whether it is probable to intersect or not.

Let us consider a simple polygon P. Now, to find the intersections in the guard zonal regions and to exclude the intersecting regions, we use line sweep algorithm [13]. Hence, while computing the initial guard zone G, we surround P by G, which is essentially a collection of $O(n)$ line segments only (and there is no circular arc part as we have drawn so far for each of the convex polygonal vertex of P); we explain it as follows taking Fig. 6 into consideration.

Now, we explain how we replace a circular arc that has been drawn so far for each of the convex polygonal vertex v of P with the help of a collection of smaller straight line segments. Next, we claim that the desired guard zone G is computed with the help of $O(n)$ straight line segments only. To show the first part under consideration, we take the help of Fig. 4 that contains a convex polygonal vertex v along with its associated polygonal edges uv and vw. Here, for the time, we do not like to know whether u and w are convex or concave polygonal vertices, as we are only interested to consider a convex polygonal vertex v, whose guard zone is to be computed comprising a constant number of (smallest possible) straight line segments instead of a circular arc centering at v with radius r. Without loss of generality, we assume that both the polygonal vertices u and w are also convex. So, what we do, we compute two straight line segments $u'v'$ and $v''w'$, where $u'v' \parallel uv$ and $u'v' = uv$, and also $v''w' \parallel vw$ and $v''w' = vw$, and the perpendicular distance between the parallel lines for both the pairs is same as r. Hence, we obtain two rectangles $uvv'u'$ and $vww'v''$, where $vv' = uu' = ww' = vv'' = r$, since $u'u$ (or $v'v$) is perpendicular to uv and $w'w$ (or $v''v$) is perpendicular to vw. So, $u'v'$ and $v''w'$ are guard zones for the polygonal edges uv and vw, respectively. Now, we compute guard zone for the polygonal vertex v as follows.

We have already told that the guard zone for polygonal vertex v is composed of a number of straight line segments that collectively replace the guard zonal circular arc (of radius r) that we usually draw at a convex polygonal vertex v outside the polygon (where $u'v'$ and $w'v''$ are two tangents to that circular arc). We like to do this task recursively using a constant time computation for each such polygonal vertex v, as the value of $\angle v'vv''$ is always less than 180°. In other words, we may state that we like to replace the circular arc that we usually compute as a part of G for v by exactly 2^p number of straight line segments (for some constant p) that are equal in length to each other.

In order to obtain the smallest possible line segments, we follow a recursive procedure which is binary in nature. First of all, we bisect $\angle v'vv''$ (or $\angle uvw$) by a bisector bs_1 whereon p_1 is a point outside the polygon such that $vp_1 = r$. We join $v'p_1$ and p_1v'', so $v'p_1v''$ is an approximated guard zone (for $p = 1$) of the circular arc we liked to draw. Next, to make this approximation finer, we further bisect $\angle v'vp_1$ by a bisector bs_2 whereon p_2 is a point outside the polygon such that $vp_2 = r$ and bisect $\angle p_1vv''$ by a bisector bs_3 whereon p_3 is a point outside the polygon such that $vp_3 = r$. Then, we join $v'p_2$, p_2p_1, p_1p_3, and p_3v'', so $v'p_2p_1p_3v''$ is a finer approximation of the guard zone (for $p = 2$) than the previous one (i.e., $v'p_1v''$) of the circular arc we usually draw.

Needless to mention that for $p = 3$, we are supposed to bisect each of the angles $\angle v'vp_2$ through $\angle p_3vv''$ and obtain intermediate points p_4 through p_7 on each such bisection bs_7 through bs_7, outside the polygon such that $vp_4 = vp_5 = vp_6 = vp_7 = r$, and even smaller line segments $v'p_4 = p_4p_2 = p_2p_5 = p_5p_1 = p_1p_6 = p_6p_3 = p_3p_7 = p_7v''$, and in due course, we obtain an even finer approximation of the guard zone (for $p = 3$) than that we computed for $p = 2$, which is more closer to the circular arc we usually draw as a part of guard zone for a convex polygonal vertex.

This process of bisection is continued till the value of each bisectional angle becomes 0.50° or 0.25° or up to some precision of angle that makes the straight line segments as chords of the circular arc reasonably very small. Now, it is very clear that all the points p_i over the bisections outside the polygon are the points on the circular arc as part of G, and each small line segment (whose length tends to zero for a smaller value of r) is an approximation of its associated arc for which it is the largest chord. So, further bisection of each of the bisected angles in subsequent levels of recursion and spotting a point p_i on each of the bisections at distance r from v outside the polygon helps to achieve more points on the said circular arc that are consecutively equidistant and closer to each other. Hence, up to some desired level of precision, we may obtain a set of equal-length straight line segments collectively that replaces the circular arc we liked to draw as a part of G for vertex v.

In any case, as the value of $\angle v'vv''$ is always less than 180°, which is a constant, we claim that the number of bisections or the number of recursive calls to bisect $\angle v'vv''$ is always a constant (up to an acceptable smallest precision of angle). Even then, this method may lack in finding a point of intersection of two circular arcs or a circular arc and an edge of G where the segments are tangential (or almost tangential) to each other. However, the probability of occurrences of such a circumstance significantly reduces as in general, the value of r is appreciably small. If some smaller edge (or the initial or final chord or some intermediate chord $p_i p_{i+1}$ that approximates its coupled arc) is found, that intersects with other part(s) of G, which is also a line segment, then we may further bisect only that angle recursively to identify a more accurate point of intersection that in time may reduce many redundant computations. In this context, we like to conclude that the total number of straight line segments as part of the computed guard zone G is at most $O(n)$, as stated below.

Now, we can compute all the intersections among the parts of the guard zone, which are now transformed into line segments instead of circular arcs. This is not a challenging problem as we can take each pair of segments, compute whether they intersect, and if so, we report their intersection points. This is a brute force approach and clearly requires $O(n^2)$ time. In some cases, it may be optimal, when each pair of line segments really intersects. Our objective is to have an algorithm that is faster in some situations where number of intersecting line segments is considerably less than total number of line segments. The line sweep algorithm [12] is such an algorithm whose running time depends not only on the number of segments in the input, but also on the number of intersection points. For this reason, this algorithm is known as output-sensitive algorithm or in this case, we may call it intersection sensitive algorithm, because the number of intersections determines the size of the output or in other words running time of the algorithm.

Proceeding in this way, we can use line sweep algorithm as all the circular arcs have been replaced by a set of only line segments, taking all the line segments which are actually in guard zone and which have been derived. But we yet do not know which pair of circular arc(s) and/or line segment(s) have been intersected. If we can somehow detect the probable intersecting regions, the number of checking can be reduced, and thus, we would apply the above procedure for a smaller set of regions which are in fact probable to intersect.

So, our first phase of the algorithm searches for the probable regions of intersections. The results achieved from this phase reduce the sample space to be checked for original intersections. In this phase, we take the extended guard zone instead of the original guard zone and that may be derived in the following way. We can take each circular arc and extend its two tangents which are actually the extended line segments of the guard zone attached to that circular arc. Thus, they meet at a point and we get s points for s number of circular arcs.

For example, an overestimated guard zone is formed by extending two neighboring guard zonal line segments drawn for the convex polygonal vertex v that meet at point p, as shown in Fig. 5. Here, the circular arc $v'v''$ is the actual guard zonal region of the convex vertex v. Thus, all the circular arcs of the guard zone are now replaced by corresponding convex vertices in the extended polygon. It may be simple or not. If the extended guard zone which is actually a polygon is simple; that is, there is no intersection between any regions of the polygon, it is sure that our original polygon has no intersections. On the other hand, if it has some intersections, those may result in actual intersections, as those are detected

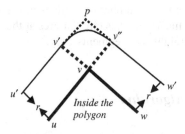

Fig. 5 Extension of the two neighboring guard zonal line segments that meet at p, instead of a circular arc of radius r, overestimates the guard zone for a convex polygonal vertex v

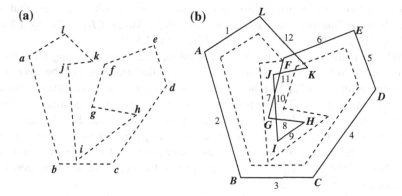

Fig. 6 a A simple polygon P drawn by *dotted line* segments. **b** An extended guard zone X of P (where X is drawn by *solid line* segments)

from the extended guard zone. So, to be sure of the intersection(s), we have to proceed further.

In the second phase, we deal with the original guard zonal regions, not with the extended guard zones but with the regions that are proved to have probable intersection(s) by the first phase of the algorithm. As has been discussed earlier, those regions involving a convex polygonal vertex are subdivided into line segments and line sweep algorithm is further applied on those line segments (computed for the guard zone) only. At the end of the second phase, we get the unique intersections and depending on this information, the algorithm reports the outer guard zone. Two phases of the algorithm is discussed below with the help of an example simple polygon, as shown in Fig. 6a.

Let us consider a simple polygon P whose vertices are stored in anticlockwise manner as a through l, where $a, b, c, d, e, f, g, k,$ and l are convex and $h, i,$ and j are concave vertices (Fig. 6a). This can be achieved by traversing the polygon in anticlockwise direction starting from a point and ending at that very point, and at the time of traversing, the vertices are checked to inform whether it is convex or concave; accordingly, this information is stored with the corresponding vertices.

This takes $O(n)$ time, if n is the number of vertices in the simple polygon. So the guard zone computed in linear time has circular arcs at the convex regions [1], and the other portions are straight line segments only.

3.1 Phase-I of the Algorithm

As has been discussed earlier, every two neighboring line segments of every circular arc (guard zone of a convex vertex) are extended and they meet at a point which is again a convex vertex of the overestimated polygon. Let us consider an overestimated guard zone X (for polygon P), whose vertices are stored in anti-clockwise manner that are A through L, as shown in Fig. 6b. The edges that are considered as line segments are labeled as $AB(2)$, $BC(3)$, $CD(4)$, $DE(5)$, $EF(6)$, $FG(7)$, $GH(8)$, $HI(9)$, $IJ(10)$, $JK(11)$, $KL(12)$, and $LA(1)$.

When we are to apply line sweep algorithm, i.e., to avoid testing all pairs of segments for intersections, we have to select lines that are close together. We need to sweep a line, parallel to a horizontal line, downward over the plane, starting from a position above all segments, and at the time of sweeping the imaginary line, we keep track of all the segments. The status of the sweep line is the set of segments intersecting it. The status is updated at particular points, not continuously. We call these particular points the *event points*. At every event point, the neighbors of the line segments are found. Only the neighbors are considered as candidates for intersection.

In the traditional line sweep algorithm [12], an assumption has been taken that no two line segments have same starting point. The neighbors of a line segment are considered depending on the intersection points of the line segments with the sweep line which is parallel to the x-axis. If no two line segments have same starting point, it is straightforward to find the neighbors at any moment. But in case of the guard zonal line segments, as a polygon is a closed region, at the top most point, it is the starting point of two line segments. At that event point, we depend on the end point to detect the order of neighborhood between the two lines.

According to the algorithm devised in this paper, we have considered the event points as the vertices, the starting and ending points of the extended guard zonal edges of the given polygon P. Here, these are stored in an event queue according to the decreasing order of y-coordinates. If two or more vertices have the same y-coordinate, then sort them according to increasing x-coordinate. So, here, the order is L, E, A, F, K, J, D, G, H, I, B, and C. These are stored in a list which is initially empty. At the starting of the algorithm, this array only contains all the starting and ending points, but subsequently, the intersection points are also inserted in it, which are later on treated as event points as well. To traverse this list efficiently and for updating at the time of inserting an intersection point, we represent this in a binary search tree.

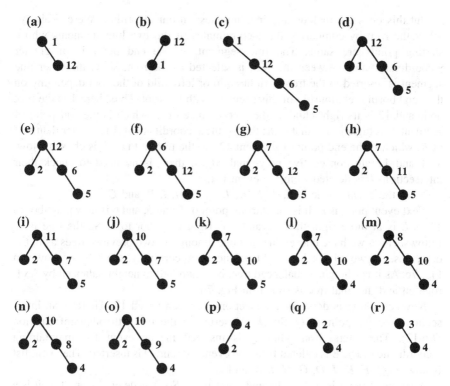

Fig. 7 **a** 1 and 12 have been inserted making 12 as the right child of 1. **b** 1 and 12 have been inserted making 1 as the left child of 12. **c** 5 and 6 have been inserted making 6 as right child of 12 and 5 as right child of 6. **d** The unbalanced tree in Fig. 7c is made height balanced. **e** 1 is deleted, and 2 is inserted making 2 as left child of 12. **f** The position of 12 and 6 has been interchanged. **g** 6 is deleted, and 7 is inserted at the position of 6. **h** 12 is deleted, and 11 is inserted at the position of 12. **i** The position of 7 and 11 has been interchanged. **j** 11 is deleted, and 10 is inserted at the position of 11. **k** The positions of 7 and 10 have been interchanged. **l** Five is deleted, and 4 is inserted at the position of 5. **m** 7 is deleted, and 8 is inserted at the position of 7. **n** The position of 8 and 10 has been interchanged. **o** 8 is deleted, and 9 is inserted at the position of 8. **p** 9 and 10 are deleted resulting 4 to be the root. **q** 9 and 10 are deleted resulting 2 to be the root, and this tree is selected as the x-coordinate of the end point of 2 is less than that of 4, though both of them have same y-coordinates. **r** 2 is deleted, and 3 is inserted at the position of 2

Again we use a Query tree T to handle the line segments which is initially empty. The lines (already labeled in clockwise or anticlockwise manner) are to be inserted or deleted or updated at the event points.

Now, the sweep line is set parallel to x-axis at the point where y-coordinate is maximum, i.e., at L. It is the starting point of two line segments 1 and 12. Now, the x-coordinate of the end of 1 is less than that of the end of 12. So, in T, 1 is left neighbor of 12. Accordingly, the tree may be any one like the trees shown in Fig. 7a, b.

But this does not include any indefiniteness in our algorithm. We can always select the root by comparing the *y*-coordinates of the two line segments whose starting points are same. The line segment, whose end point is of greater *y*-coordinate value between the two, is selected as the root. Now, the other line segment is inserted in the tree as right child or left child of the root depending on the neighboring relationship of this segment with the root. Thus, here 1 is the root node and 12 is its right child as the *y*-coordinate of *A*, which is the end point of segment 1, is higher than that of *K*; that is, the *y*-coordinate of *A* is greater than that of *K*, which is the end point of segment 12. So the tree in Fig. 7b is chosen. Now, as 1 and 12 are consecutive polygonal edges, there is no need to check their intersection. *L* is deleted from the event list.

Now, the event queue is *E*, *A*, *F*, *K*, *J*, *D*, *G*, *H*, *I*, *B*, and *C*.

Next event point is *E*. It is the starting point of 6 and 5, and 6 is now neighbor to 12 as *E* is at more right than 12 and 5 is right neighbor to 6. So the tree is as follows. Here, we have chosen the first one among the two previous trees as in the other case, it would not be height balanced. Accordingly, the tree is shown in Fig. 7c. As it is a height-imbalanced tree, it is modified to height balanced by AVL rotation and the final tree is shown in Fig. 7d.

Now, checking is done for intersections between 6 and 12. There is an intersection, and the point is *Q*. So *Q* is inserted in the list with other information *Q*(6,12). The intersection point *Q* is inserted in the event list maintaining *y*-coordinate order. *E* is deleted from the event list, and *Q* is inserted. The event list is now *A*, *Q*, *F*, *K*, *J*, *D*, *G*, *H*, *I*, *B*, and *C*.

Next event point is *A*. It is the end point for 1. So, 1 is deleted from *T* as it is a leaf node; nothing to do for updating. Now, 2 is inserted in *T* and it is the left neighbor of 12 as 1 has been deleted and *A* is at left of the point at which 12 cuts the sweep line. Accordingly, the tree is shown in Fig. 7e. *A* is deleted from the event list. Now, the event list is *Q*, *F*, *K*, *J*, *D*, *G*, *H*, *I*, *B*, and *C*.

Now, the event point *Q* is to be handled. As it is an intersection point, the neighboring information is updated in *T*. Twelve and 6 are interchanged. Again their set of neighbors has also been interchanged. Accordingly, the tree is shown in Fig. 7f. *Q* is deleted from the event list. The event list is obtained as *F*, *K*, *J*, *D*, *G*, *H*, *I*, *B*, and *C*.

Next event point is *F*. It is the end point of 6 and starting point of 7. Line 6 is deleted from *T*, and line 7 is inserted into *T*. As *F* is at left of 12 but right of 2, 12 is the right neighbor and 2 is the left neighbor of 7. So the tree is shown in Fig. 7g. *F* is deleted from the event list. The event list is now *K*, *J*, *D*, *G*, *H*, *I*, *B*, and *C*. Now, 2 and 7, and 12 and 7 are to be checked for intersection. There is no intersection.

Next event point is *K*, and it is the ending point of 12 and starting of 11. Now, *K* is at the right of 7 and left of 5; so the left neighbor and right neighbor of 11 are 7 and 5, respectively. Next, 12 is deleted and 11 is inserted. Accordingly, the tree is shown in Fig. 7h. Now checking for intersection is done between lines 7 and 11, and lines 5 and 11. Now, there is intersection between 11 and 7, and the point is *p*. So the intersection list is updated *q*(6, 12), *p*(7, 12). *p* is inserted in the event list, and *K* is deleted from it. The event list is now *P*, *J*, *D*, *G*, *H*, *I*, *B*, and *C*.

Next event point is p. As it is an intersection point, the neighboring information is updated in T. Eleven and 7 are interchanged. Again their set of neighbors has also been interchanged. Accordingly, the tree is shown in Fig. 7i. p is deleted from the event list. The event list is now J, D, G, H, I, B, and C.

The next event point is J. It is the end point of 11 and starting of 10. Now, J is at the right of 2 and left of 7. So, the left neighbor and right neighbor of 10 are 2 and 7, respectively. Eleven is deleted, and 10 is inserted. Hence, the tree obtained is shown in Fig. 7j. Now, 10 and 7, and 10 and 2 are to be checked for intersection. There is an intersection between 10 and 7, and the point is R. So the intersection list is updated $q(6, 12)$, $p(7, 12)$, $r(11, 7)$. r is inserted in event list, and J is deleted from the event list. The event list is now r, D, G, H, I, B, and C.

Next event point is r. As it is an intersection point, the neighboring information is updated in T. Ten and 7 are interchanged. Again their neighbor sets have also been interchanged. The resulting tree is shown in Fig. 7k. Now, r is deleted from the event list. The event list is now D, G, H, I, B, and C.

Next event point is D. It is the end point of 5 and starting of 4. Now, D is at the right of 10. So the left neighbor of line 4 is line 10. Five is deleted, and 4 is inserted. Accordingly, the tree is shown in Fig. 7l. Now checking for intersection is done between 10 and 4. There is no intersection. D is deleted from the event list. The event list is now G, H, I, B, and C.

Next event point is G. It is the end point of 7 and starting of 8. Now, G is at the right of 2 and left of 10. So, the left neighbor and right neighbor of 8 are 2 and 10, respectively. Seven is deleted, and 8 is inserted. The tree is shown in Fig. 7m.

Now, checking for intersection is done between 10 and 8, and 2 and 8. There is an intersection between 10 and 8, and the point is s. Hence, the intersection list is updated as $q(6, 12)$, $p(7, 12)$, $r(11, 7)$. $s(10, 8)$. s is inserted in event list, and G is deleted from the event list. The event list is now S, H, I, B, and C.

Next event point is s. As it is an intersection point, the neighboring information is updated in T. Ten and 8 are interchanged. Again their neighbor sets have also been interchanged. Accordingly, the tree is shown in Fig. 7n. s is deleted from the event list. The event list is now H, I, B, and C.

Next event point is H. It is the end point of 8 and starting of 9. Now, H is at the right of 10 and left of 4. So, the left neighbor and right neighbor of 9 are 10 and 4, respectively. Eight is deleted, and 9 is inserted. The tree is shown in Fig. 7o. Now, 9 and 10, and 9 and 4 are to be checked for intersection. There is no intersection, so H is deleted from the event list. The event list is now reduced to I, B, and C.

Next event point is I. It is the end point of 10 and 9. Now, I is at the right of 2 and left of 4. So, 9 and 10 are deleted. Accordingly, the tree is shown in Fig. 7p. Now, checking for intersection is done between 2 and 4. There is no intersection, so I is deleted from the event list. The event list is now B and C.

Next event point is B. It is the end point of 2 and starting of 3. B is at the left of 4. So, 3 is inserted at left of 4, and 2 is deleted from T. Hence, the tree we obtain is shown in Fig. 7q. No checking is made, as 3 and 4 are two consecutive edges of the polygon. B is deleted from the event list, and the event list we obtain is C.

So, the next event point is C. It is the end of 3 and 4. So, 3 and 4 are deleted, and the tree T as well as the event list becomes empty.

The intersection list is now considered. For each intersection detected in the above phase, a second phase of the algorithm comes into picture.

3.2 Phase-II of the Algorithm

Now, we have already obtained the intersection points for the overestimated guard zonal region. The intersection point can be on the actual guard zonal line segment or on the extended portion of the line segment which has been drawn to make the guard zone a polygon, which is not simple in general. So, for each intersection point, we have to check among the original line segment and the circular arc(s) depending on the fact whether the line segment joins two convex vertices, or one concave vertex and one convex vertex. If it joins two concave vertices, then only the line segment is considered.

If we repeatedly bisect the obtained angles after some iterations when the angles are getting much smaller (with respect to some predefined value), the arcs can be considered as line segments. Even when the convex angle tends to $360°$, we have to subdivide only $180°$ as we exclude two $90°$ angles before starting the subdivision.

If we subdivide the arc p times, there are $p + 1$-ordered subarcs or line segments where two consecutive line segments share their starting and ending points except the two at the end. So, the number of event points is $p + 2$ for each circular arc.

In the intersection list, the first intersection point is q, which is obtained in between lines 6 and 12. Line 6 joins two convex points F and E. So, we have to consider both the circular arcs corresponding to E and F, and the line segment joining those circular arcs. Again, line 12 also joins two convex points L and K, and we consider both the circular arcs corresponding to L and K, and the line segment joining these circular arcs.

From the above information, we have four circular arcs and two line segments for a probable intersecting region which we have got from the first phase of the algorithm. At the second phase, after subdividing the arcs, we get a total of $4(p + 1) + 2$ number of line segments for the line sweep algorithm. After applying the line sweep algorithm on the line segments mentioned above, we find that actually there is no intersection among the original guard zone.

The next intersection point is p, which is obtained in between lines 7 and 11. Line 7 joins two convex points F and G. So, we have to consider both the circular arcs corresponding to F and G, and the line segment joining these circular arcs. Again 11 joins a convex point K and a concave point J; so we consider the circular arc corresponding to K and the line segment joining the circular arc and the concave point J.

Now, we have three circular arcs and two line segments for a probable intersecting region. At the second phase, after subdividing the arcs, we get a total of

$3(p + 1) + 2$ number of line segments for the line sweep algorithm. After applying the line sweep algorithm on the line segments mentioned above, we find that actually there is also no intersection among the original guard zone.

From the intersection list, our next intersection point we get is r, which is formed in between lines 7 and 10. Line 7 joins two convex points F and G. So, we consider both the circular arcs corresponding to F and G, and the line segment joining these circular arcs. Again, line 10 joins two concave points J and I, so we consider only the line segment joining the concave points J and I.

Now, we have two circular arcs and two line segments for a probable intersecting region. At the second phase, after subdividing the arcs, we get a total of $2(p + 1) + 2$ number of line segments for the line sweep algorithm. After applying the line sweep algorithm on the line segments mentioned above, we find that there is one intersection point between the circular arc corresponding to G and the line segment JI. Let the point be Y. We store the intersection point as a triple ⟨intersection point, two ends of one of the intersecting line, two ends of the other intersecting line⟩.

If the line joins two circular arcs, then the two ends of the line segment are denoted as (arc, arc). If the line joins two concave points, the ends are defined as (end point, end point). If the line joins one circular arc and one concave point, then the ends are defined as (arc, end point). Again we store the ending information of that very line which first contains the end point or arc which occurs before others in the list of the line segments and arcs of the guard zone. So, here we store ⟨Y, (arc($F1$, $F2$), arc($G1$, $G2$)), (I, J)⟩.

From the intersection list, our next intersection point is s, which is formed in between lines 8 and 10. Eight joins one convex point G and one concave point H. So we consider the circular arcs corresponding to G and the line segment joining the circular arc and the concave point H. Again, 10 joins two concave points J and I, so we consider only the line segment joining the concave points J and I.

Now, we have one circular arc and two line segments for a probable intersecting region. At the second phase, after subdividing the arcs, we get a total of $(p + 1) + 2$ number of line segments for line sweep algorithm. After applying the line sweep algorithm on the above-stated line segments, we find that actually there is one intersection between the line segment joining the circular arc corresponding to G and the concave point H and the segment JI. Let the point be X. We store it as ⟨X, (arc($G1$, $G2$), H), (I, J)⟩.

The output is in the form of the list of guard zonal line segments and guard zonal circular segments after eliminating intersecting region. As in the notch area, guard zone may be divided into inner and outer guard zone after eliminating the overlapping region. After detecting the intersecting points, they are also updated in the list. Thus, before applying this algorithm, our list is in the form (Fig. 8):

arc($A1$, $A2$), $A2B1$, arc($B1$, $B2$), $B2C1$, arc($C1$, $C2$), $C2D1$, arc($D1$, $D2$), $D2E1$, arc($E1$, $E2$), $E2F1$, arc($F1$, $F2$), $F2G1$, arc($G1$, $G2$), $G2H$, HI, IJ, $JK1$, arc($K1$, $K2$), $K2L1$, arc($L1$, $L2$), $L2A1$.

Fig. 8 The computed entire
guard zone of the given
polygon *P* in Fig. 6a

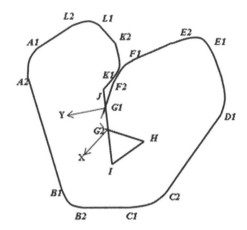

As our objective is to determine the intersection and update the original guard zone eliminating it, we focus only on the intersection points. We start our stored list of the line segments and circular arcs and traverse the guard zone anticlockwise. When one of the intersection points is achieved (say, this is the intersection between $L1$ and $L2$ and we are traversing through $L1$), we do not proceed through the line($L1$) we were so far traversing. We change the direction and continue traversing through the other line segment or arc($L2$) intersected at that point. From the intersection point, we change our path, and through the other line, the traversing is continued in anticlockwise manner. We can find the next line segment or arc after $L2$ in anticlockwise direction, as the original guard zone list is already stored. Thus, when we reach at the starting point of the traversal, our job is done. It is performed in $O(n)$ time if the number of vertices in the polygon is n. In our example, it results in the outer guard zone as follows: arc($A1$, $A2$), $A2B1$, arc($B1$, $B2$), $B2C1$, arc($C1$, $C2$), $C2D1$, arc($D1$, $D2$), $D2E1$, arc($E1$, $E2$), $E2F1$, arc($F1$, $F2$), $F2Y$, YJ, $JK1$, arc($K1$, $K2$), $K2L1$, arc($L1$, $L2$), $L2A1$.

In this case, when we arrive at Y after traversing segment $F2Y$, we check for the line segment which intersects at Y other than $F1G2$. Here, it is JI and the vertex from Y at anticlockwise direction is J. So, we move to J and report the segment YJ as the next traversed line segment in the original guard zone excluding the overlapped regions.

Sometimes, there may be an overlapping at the notch region and there is a sufficient place in that notch to place a subcircuit to utilize the area more efficiently. In this case, if we follow the above procedure, we compromise the possibility to find the region which is in the shape of a loop inside the notch. In that case, we follow the procedure said below.

We start traversing the guard zonal line segments and circular arcs as said above. When we are at one of the intersection point, we traverse anticlockwise enlisting the line segments and circular segments including the intersection point also. Thus, the list starting from one intersection point and ending at the same point is to be eliminated from the guard zone as it includes the inner guard zone and intersection region. Then, we get the resultant list for the guard zone. The inner guard zone can also be specified by sublist of the above-said list. If there is one such cycle starting from one intersection point and ending at that point without having any other intersection point within it, it is the inner guard zone.

So, from our example, we have got two intersection triples
$\langle Y, (\text{arc } (F1, F2), \text{arc } (G1, G2)), (I, J)\rangle$
$\langle X, (\text{arc } (G1, G2), H), (I, J)\rangle$.
Starting from arc $(F1, F2)$, we get the list:
$\text{arc}(F1, F2)$, $F2Y$, $YG1$, $\text{arc}(G1, G2)$, $G2X$, XH, HI, IX, XY, YJ as it covers all the end points of this triple and it is updated in the original guard zonal list. But here is no inner guard zone starting from Y and ending at Y because in this cycle, there is another intersection point X. Before updating the original list, we remove the sublist starting from Y and ending at the line segment joining two intersection points. Thus, here we remove this portion: $YG1$, arc $(G1, G2)$, $G2X$, XH, HI, IX, XY.
Starting from arc $(G1, G2)$, we get the list: arc $(G1, G2)$, $G2X$, XH, HI, IX, XY, YJ
So the inner guard zone is XH, HI, IX.
The outer guard zone is $\text{arc}(A1, A2)$, $A2B1$, $\text{arc}(B1, B2)$, $B2C1$, $\text{arc}(C1, C2)$, $C2D1$, $\text{arc}(D1, D2)$, $D2E1$, $\text{arc}(E1, E2)$, $E2F1$, arc $(F1, F2)$, $F2Y$, YJ, $JK1$, $\text{arc}(K1, K2)$, $K2L1$, $\text{arc}(L1, L2)$, $L2A1$.

3.3 Algorithm at a Glance

The first phase of the algorithm to detect the probable regions of intersections:

Input: Vertices of the polygon (where we have considered the guard zone) in anticlockwise manner. The edges, which are considered as line segments here, are labeled.

Event points: Vertices (the starting and ending points of the polygonal edges) intersection. An event queue is maintained to store the event points, and after traversing a point, it is deleted from the event queue.

Query tree (T): the line segments (already labeled in clockwise or anticlockwise manner) are to be inserted or deleted or updated at the event points in a tree structure. Initially, the tree is empty.

Step 1: Sort all the vertices in the decreasing order of y. If two or more vertices have the same y-coordinate, then sort them according to increasing x-coordinate.

Step 2: Set the sweep line parallel to x-axis at the point where y-coordinate is maximum. It is the first event point.

Step 3: At any event point, update the query tree T if it is a starting point.

Step 3.1: Insert the starting line segments to the query tree (T) according to the order of the x-coordinates of the event points. As there may be two starting lines at an event point, they are to be inserted in the query tree having proper position with respect to x-coordinate of the end points of those line segments.

Step 3.2: Then, check for intersection with those who are neighbors of the line segment in the query tree. No checking is required for those pair of neighbors which are consecutive edges in the polygon. If there is an intersection, then retrieve the information as follows.

Step 3.2.1: Intersection point is stored in the intersection list as *intersection point*; the line segments intersected at that point.

Step 3.2.2: The intersection point is inserted in the event queue at the position maintaining y-coordinate order.

Step 3.3: If it is an ending point, then delete the line and update the neighboring nodes. The previous neighbors of the deleted line are now neighbor of the new line.

Step 3.4: If it is an intersection point, update the query tree T by interchanging the positions of the intersecting lines.

Step 3.5: Delete the event point from the event queue.

Step 3.6: Terminate when the entire event points are visited; that is, event queue is empty and the query tree T is again empty.

Output: The intersection list of which every entry is in the following form: $s\langle v_i(l_1, l_2)\rangle$ v_i denotes the intersection point, (l_1, l_2) denotes (line 1, line 2 that have intersected at that point v_i).

The second phase of the algorithm to detect the actual intersections from probable regions of intersections
Input:

- The intersection list obtained from Phase-I.
- Corresponding line segments and arcs of the original guard zone of the given polygon to the vertices of the extended polygon which were enlisted in the intersection list.

Step 1: For every intersection point, two lines intersecting at that point are considered and the list of the ends of those two line segments is built.

Step 1.1: If the two lines join two circular arcs each at the ends, the list contains four circular arcs and two line segments which intersect.

Step 1.2: If both the two lines join two concave points each, the list contains only two line segments which intersect.

Step 1.3: If one of the lines joins one circular arc and one concave point and another joins two circular arcs, then the list contains three arcs and two line segments which intersect.

Step 1.4: If both the lines join one circular arc and one concave point each, then the list contains two arcs and two line segments which intersect.

Step 2: The following is performed for each circular arc.

Step 2.1: Every circular arc is the guard zonal region for one convex vertex of the corresponding original polygon. Two perpendiculars are drawn on the two points where the circular arcs meet the neighboring line segments.

Step 2.2: The angle between these two perpendiculars is bisected. Thus, we get two subarcs of equal size. Then, every subarc is bisected again. This is a recursive procedure, and $p + 1$ subarcs are there after bisecting p times. p is predefined, such that $p + 1$ subarcs of equal size can be obtained.

Step 2.3: p is so chosen that the subarcs obtained are so small that they can be treated as line segments. So, there is $p + 1$ line segments for each arc which shares p number of points as common point of every two consecutive line segments. If the list as said above contains 'x' number of circular arcs, it has now $(x*(p + 1)) + 2$ line segments.

Step 3: Now, line sweep algorithm is applied to the line segments of this list as described in the first phase of the algorithm. As the information of consecutiveness among the lines is known, if there exists an intersection point except these common points, it is stored as it is the intersection point between the original guard zone regions.

Thus, before applying this algorithm, the list contains all the line segments and arcs without considering the intersection regions. Whenever the original intersection points are obtained (if any), the list is updated by inserting the intersection point(s) on the line segment(s) and/or arc(s) and renaming the line segments or arcs by dividing it at the intersection point(s).

At the time of reporting the guard zone excluding the overlapped regions, starting from the end point as listed in one of the intersection triple, the polygon is traversed anticlockwise enlisting the line segments and circular segments including the intersection points also. Thus, the list starting from one intersection point and ending at the same point is to be eliminated from the guard zone as it includes the inner guard zone and intersection region. The resultant list for the guard zone is thus obtained. The inner guard zone can also be specified by sublist of the above-said list. If there is one such cycle starting from one intersection point and ending at that point without having any other intersection point within it, it is the inner guard zone.

After finding list of line segments and arcs, it is updated in the original guard zonal list. There is no inner guard zone starting from one intersection point (Y) and returning to that very intersection point (Y) if in this cycle, there is not any other intersection point X and $X \neq Y$; otherwise, there is an inner guard zone which is to be distinctly specified. If there exists any inner guard zone, before updating the original list, the sublist starting from Y and ending at the line segment joining two intersection points X, Y is removed. The remaining list is the outer guard zone.

4 Complexity Analysis

If the number of edges in the original polygon is n, then the number of edges in the overestimated polygon is also n. The algorithm starts by constructing the event queue by sorting the starting and end points of the line segments, which takes $O(n \log n)$ time. Initializing the status structure takes constant time. The handling of event queue consists of three operations, insertion, deletion, and interchange of positions, which takes $O(\log n)$ time each. Now, $m = n + I$; I is the number of intersection points. The complexity of line sweep algorithm is $O(m \log_2 n)$ [13].

Lemma *The number of times any circular arc is to be subdivided to convert one of the division a line segment, i.e., p is a constant.*

Proof We consider two extreme cases, i.e., the circular arcs for the convex vertices with two extreme values of the angles. If we can prove that p is a constant for those two cases, it is true for all the intermediate values. The convex angle is maximum when it tends to 180° and it is minimum when it tends to 0°. Figure 9a, b show these two cases. If internal angle tends to 180°, the external angle tends to 180°. Hence we draw perpendiculars from the vertex of the original polygon to the two adjacent line segments at points A and B as shown in Fig. 9a. So, $\angle AOB$ tends to 0° and we get the line segment corresponding to the circular arc by connecting A and B. Thus there is no need to bisect the angle for this situation.

On the other hand if the external angle tends to 360°, observing Fig. 9b, $\angle AOB$ which tends to $360° - (90° + 90°) = 180°$, is to be subdivided. When a sub arc makes an angle less than or equal to 4°, we can consider the sub arc to be a line segment. To subdivide the $\angle AOB$ here, if we follow the sequential algorithm, we need to perform the bisection operations 44 times, which is a constant, i.e., it does not depend on n, the number of vertices in the original polygon. Hence p is a constant between 0 and 44. ◆

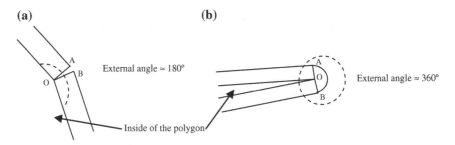

(a)   **(b)**

External angle ≈ 180°

Inside of the polygon

External angle ≈ 360°

Fig. 9 **a** External angle tends to 180°, but the circular arc to be subdivided is associated with an angle that tends to 0°. **b** External angle tends to 360°, but the circular arc to be subdivided is associated with an angle that tends to 180°

Again, if the number of intersections is I, then the maximum number of line segments that take part in the line sweep algorithm in the second phase is $(4(p + 1) + 2)$ for each case, where p is the number of iterations by which the bisection has been done. So, the complexity is $O(p \log_2 p)$. For I intersection points, it is $O(Ip \log_2 p)$. For two phases, it is $O(n \log_2 n + Ip \log_2 p \log_2 n)$. As p is a constant predefined, $p \log_2 p$ is also a constant; we can conclude it as $O(n \log_2 n + cI \log_2 n)$.

The amount of storage used by the algorithm is to be analyzed. In the first phase of the algorithm, the tree T stores a segment at most ones, so it uses $O(n)$ storage. The size of the event queue is bounded by $O(n + I)$ [12]. In the second phase, we need constant (in terms of p) amount of storage for every intersection point detected in Phase-I, ultimately resulting in linear time.

5 Applications and Conclusion

Now, in brief, we like to point out the importance and motivation of the problem as follows. Suppose, there are two (approximated) guard zones G_1 and G_2 that are computed for two 2D simple polygons P_1 and P_2, respectively, those are not shown in Fig. 10a. Moreover, these two polygons are to be placed adjacent in realizing a larger VLSI circuit, where the two polygons or guard zones must not overlap. So, there might have several 2D arrangement (or placement) of these two guard zones as shown in Fig. 10b–e, out of which the placement in Fig. 10d takes the most reduced space (or area).

Though we have considered here a simple polygon, sometimes there may be more than one subcircuits whose guard zonal regions are somewhere so close that

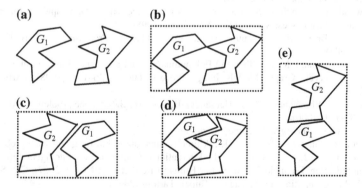

Fig. 10 **a** Two (approximated) guard zones G_1 and G_2 are assumed as computed for two 2D simple polygons P_1 and P_2 (that are not shown in these figures), respectively. **b–e** Different 2D arrangement (or placement) of these two guard zones, out of which (**d**) consumes the least amount of 2D space (due to better use of notches)

they overlap. This compels us to compute a common guard zonal region for them removing the intersection regions.

It may so happen that sometimes a small polygon that has been placed outside a large polygon with a sufficiently big notch in it. In this case, the small polygon could be accommodated inside the notch of the large polygonal boundary. Often, this sort of placement of a small polygon inside a notch of some other polygon may provide a compact design and subsequently, space is also saved. Thus, resizing is an important problem in VLSI layout design as well as in embedded system design, while accommodating the (groups of) circuit components on a chip floor, and this problem motivates us to compute a guard zone of a simple polygon.

The guard zone problem finds another important application in the automatic monitoring of metal-cutting tools. Here, a metal sheet is given and the problem is to cut a polygonal region of some specified shape from that sheet. The cutter is like a ballpoint pen whose tip is a small ball of radius δ, and it is monitored by a software program. If the holes inside the notch also need to be cut, our algorithm can easily be tailored to satisfy that requirement too.

The Minkowski sum is an essential tool for computing the free configuration space of translating a polygonal robot [1]. It also finds application in the polygon containment problem and in computing the buffer zone in geographical information systems [5], to name only a few.

References

1. Bajaj, C., Kim, M.-S.: Generation of configuration space obstacles: The case of a moving algebraic curves. Algorithmica **4**(2), 157–172 (1989)
2. de Berg, M., van Kreveld, M., Overmars, M., Schwarzkopf, O.: Computational Geometry: Algorithms and Applications. Springer, Berlin (1997)
3. Hernandez-Barrera, A.: Computing the minkowski sum of monotone polygons. IEICE Trans. Inf. Syst. **E80-D**(2), 218–222 (1996)
4. Heywood, I., Cornelius, S., Carver, S.: An Introduction to Geographical Information Systems. Addison Wesley Longman, New York (1998)
5. Hwang, K., Briggs, F.A.: Computer Architecture and Parallel Processing. McGraw-Hill, New York (1984)
6. Lee, I.-K., Kimand, M.-S., Elber, G.: Polynomial/rational approximation of minkowski sum boundary curves. Graph. Models Image Process. **60**(2), 136–165 (1998). (Article No: IP970464)
7. Nandy, S.C., Bhattacharya, B.B., Hernandez-Barrera, A.: Safety zone problem. J. Algorithms **37**, 538–569 (2000)
8. http://en.wikipedia.org/wiki/Curve_orientation
9. Mehera, R., Chatterjee, S., Pal, R.K.: A time-optimal algorithm for guard zone problem. In: Proceedings of 22nd IEEE Region 10 International Conference on Intelligent Information Communication Technologies for Better Human Life (IEEE TENCON 2007), CD: Session: ThCP-P.2 (Computing) (Four pages). Taipei, Taiwan (2007)
10. Mehera, R., Pal, R.K.: A cost-optimal algorithm for guard zone problem. In: Proceedings of 10th International Conference on Distributed Computing and Networking (ICDCN 2009), pp. 91–98. Hyderabad, India (2009)

11. Mehera, R., Chatterjee, S., Pal, R.K.: Yet another linear time algorithm for guard zone problem. Icfai J. Comput. Sci. **II**(3), 14–23 (2008)
12. Goodrich, M.T.: Intersecting line segments in parallel with an output-sensitive number of processors. Soc. Ind. Appl. Math. **20**(4), 737–755 (1991)
13. de Berg, M., van Kreveld, M., Overmars, M., Schwarzkopf, O.: Computational Geometry: Algorithms and Applications. Springer, Berlin (1997)

Author Index

© Springer India 2015   211
R. Chaki et al. (eds.), *Applied Computation and Security Systems*, Advances in Intelligent
Systems and Computing 305, DOI 10.1007/978-81-322-1988-0